Environmental Effects
of Complex
River Development

Other Titles in This Series

Westview Special Studies in Natural Resources and Energy Management

Environmental Effects of Complex River Development:
International Experience
edited by Gilbert F. White

In our critical attempts to solve the pressing current problems of a limited water supply, it is essential that we act always with a global view to the future. Recognizing this, an international group of scholars—from the Soviet Union, Canada, Africa, and the United States—met to review together their experiences and research on the environmental effects of a number of large scale river management programs. This edited collection of their reports provides a balanced view of a vital element in the total ecosystem. Their analysis points out the urgent need to take account of long-term trends in climate, to consider all feasible management alternatives, and, especially, to manage demand (as opposed to simply increasing water supply) and to defer irreversible action until all environmental impacts are estimated.

Gilbert F. White, chairman of the International Geographical Union's Commission on Man and the Environment, is professor of geography and director of the Institute of Behavioral Science at the University of Colorado.

Published with the cooperation of the International Geographical Union and UNESCO.

Environmental Effects of Complex River Development

edited by Gilbert F. White

Routledge
Taylor & Francis Group

LONDON AND NEW YORK

First published 1977 by Westview Press

Published 2018 by Routledge
52 Vanderbilt Avenue, New York, NY 10017
2 Park Square, Milton Park, Abingdon, Oxon OX14 4RN

Routledge is an imprint of the Taylor & Francis Group, an informa business

Library of Congress Cataloging in Publication Data

Main Entry under title:
Environmental Effects of complex river development.
  (Westview special studies in natural resources and energy management)
  1. Water resources development—Environmental aspects—Congresses.
2. Rivers—Congresses. I. White, Gilbert
Fowler, 1911- II. International Geographical Union.
TC401.E58   333.9'102   77-3943

ISBN 13: 978-0-367-01772-9 (hbk)
ISBN 13: 978-0-367-16759-2 (pbk)

Contents

Tables

Figures

Acknowledgements

The analysis of complex river development reported in this volume grows out of international collaboration among geographers concerned with the environmental effects of large-scale transformations in the water regime of the continents. The cooperative effort was set in motion by the International Geographical Union's Commission on Man and Environment, composed of David H. K. Amiran, Hebrew University, Jerusalem; Ian Burton, University of Toronto; Inokenty P. Gerasimov, Institute of Geography, Soviet Academy of Science; Sanislaw Leszczycki, Institute of Geography, Polish Academy of Science; T. Nakano, Tokyo Municipal University; and myself.

A set of reports of complex river development was presented and reviewed at a symposium organized under the leadership of Gerasimov on the Volga-Don rivers in July 1976. Other participants in the critical review were Otmara Avello, Povilas Balsarevitchyus, Vilen Barskiy, Natalia Basilevitch, Angel Bassals-Batalla, Duane Baumann, Jacquelyn Beyer, Alexandr M. Bronfman, Ian Burton, John C. Day, Jean Dresch, Dushan Dukich, Daniel Dworkin, Fillmore Earney, Tamara Fedoseeva, J. Keith Fraser, Maria Gerasimova, James Gibson, Maria Glazovskaya, Sergei Gorshkov, Nicholas Helburn, Richard Jackson, Mary Just, Vitold Kaminsky, Robert W. Kates, David Kromm, Nikolay Ladeyshikov, Kenneth Langran, Robert Layton, Yves Masurel, Phillip Micklin, Evgenka Minaeva, Kirill Mishev, Nora Mitchell, Kazou Mitsui, Keith W. Muckleston, Liesa Nestmann, Anatolii Nikanorov, R. S. Odingo, Afulabi Ojo, Ivan Ponomarenko, Petr Popov, Philip Pryde, Hans Richter, Nikolay Rodzyanko, Robert Saveland, Leonid N. Shapiro, Robert Tamsma, Ygor Trutnev, Lem Valesyan, Semion L. Vendrov, Ernst Weigt, Jesse Wheeler, Gilbert F. White, Anne U. White, Bruce Wood, Craig ZumBrunnen, and Tatjana Zvonkova.

The papers in this volume are drawn from the experience of those who participated. W. R. Derrick Sewell, who was unable to attend, has also contributed a chapter. All of the authors have benefitted from the discussions of that group as it alternated between shipboard symposia and shore observations between Rostov on Don and Kazan on the Volga.

Publication of the resulting volume is made possible by a grant from the United Nations Educational, Scientific, and Cultural Organization. UNESCO also contributed to travel costs for participation by scientists from developing countries. Its interest and support is gratefully acknowledged.

Gilbert F. White
Boulder, Colorado

Acknowledgements

1

Comparative Analysis of Complex River Development

Gilbert F. White

Management of the world's water resource is undergoing a momentous shift. New approaches to complex river development slowly are according greater recognition to its environmental limits and consequences. These approaches are departures from the preoccupation with single-purpose water development at the beginning of the twentieth century and are radical extensions of the concept of integrated or complex river development that had gained acceptance by midcentury. Complex utilization means the integrated management of river flow and quality to serve multiple purposes. Thus, it is not limited to river basin management. It may well include segments of territory, such as metropolitan areas or a national plan (as in Kenya), cutting across several river basins. It may involve a combination of parts of two or more basins in systems of diversion and management which incorporate actions throughout a larger area, as in the proposed transfer of water from north-flowing to south-flowing streams in the USSR.

The new emphasis reflects, in part, the sobering findings of investigations of what happened in the trail of the enthusiastic

Gilbert F. White is professor of geography and director of the Institute of Behavioral Science at the University of Colorado, Boulder. This chapter draws heavily upon the discussion and findings of the International Geographical Union symposium held on the Volga-Don, July 15-25, 1976.

commitment to water development in earlier decades. In turn, it makes new demands upon scientific research. Heavier emphasis is placed upon research which cuts across the conventional disciplinary lines of engineering and the physical, social, and biological sciences to find answers to newly phrased questions relating to the environmental effects of human interventions in the hydrologic cycle. Among those questions which geographic analysis helps illuminate are the problems of anticipating long-term fluctuations in water supply, of choosing among a theoretically wide range of options, and of assessing environmental impacts.

Single-Purpose, Multiple-Purpose, and Beyond

At the beginning of the twentieth century massive engineering projects to manage water for single purposes commanded public attention and huge resources of capital and labor. The building of the Suez Canal for navigation was rivaled in number of laborers employed and cubic meters of dirt moved only by the Tiza flood control project in the Great Central Plain of Hungary. They were the greatest engineering ventures of their time. Large new irrigation projects were launched for dry lands in the Indian subcontinent and the western United States. These monumental projects generated enthusiasm at the prospect that they would yield generous returns in economic productivity. They promised to reshape the use of continental resources through altering transportation routes, the availability of water, and the accessibility of flood-free land. Indeed, they did, and the desirability of the transformations was assessed almost wholly in terms of their direct contributions to transport and agriculture.

By the end of the first decade in the century interest turned to the management of these engineering works in entire river basins. Plans for flood control in the entire Tiza or the Miami basin of Ohio began to receive special support. These efforts were supplemented and broadly expanded by the development of techniques for water projects which combined more than one purpose in a single engineering structure and which in many instances combined water supply or flood protection with the generation of hydroelectric power, thus adding a highly vendible and profitable component to the construction of impoundment works. Large dams, as symbolized in the early works at Aswan on the Nile, in the Indus barrages, and on the Colorado River became symbols of efficient application of engineering techniques to water management.[1] The vision of combining groups of multipurpose projects in entire river basins was achieved first in the Tennessee Valley in the 1930s. It was pursued but never fully captured in many other basins around the world, including the Rhone, and the Damodar in India. The Volga in the USSR, and the Snowy Mountains scheme of Australia did approach fully integrated development of an entire basin.

In the next two decades a gradual process of reevaluation and appraisal began in which accurate account was taken of the shortfalls of many of the ambitious plans for river control, and in which engineering works were examined in the framework of economic efficiency, using analytical techniques from the field of welfare economics. The effect of application of more rigorous modes of benefit-cost and cost effectiveness analysis was to raise questions as to the efficacy of some of the works already constructed or on the drawing boards. It did not, however, initially question the basic reliance upon engineering works as a means of gaining more effective use of water by storage, channelization, transfer, drainage, and treatment.

By the 1970s the integrated river development approach was undergoing further modification in four important directions. First, increased attention was given to the variety of ways in which natural resources might be developed by employing both engineering and social modes of management. Second, much greater recognition was given to the consequences for the biological and physical environment of the construction and management of engineering projects. Third, problems of water quality resulting from pollution, both from nonpoint and point sources, had attracted anxious examination. Fourth, increasing attention was paid to water management within metropolitan regions or in areas of common groundwater resource in contrast to their investigation within the more conventional river basin area.

These shifts in approaches to river development also grow out of a deepening understanding of the role of water resources development on the world scene. Perhaps the most important single effort to improve the quality of data affecting water management came in the International Hydrologic Decade (IHD) which was organized under the United Nations Educational Scientific and Cultural Organization (UNESCO) in conjunction with other specialized agencies and the International Council of Scientific Unions. Although somewhat desperate and disjointed in its enlistment of collaboration in member countries, the IHD brought together a large volume of fugitive data, encouraged the collection of new data, and stimulated analysis of experience in individual river basins and on the world scale. From its publications it is now possible to gain deeper insight into the physical, chemical, biological, and geomorphic characteristics of streams and their use. Major steps have been taken in constructing estimates of the world water balance.[2] As a result, it became possible for the first time to put together a coherent estimate of the total supply of salt and fresh water on the globe and of the way in which fresh water is distributed over the continents and used for human purposes. However, global summaries of the amount of water, its quality, and its present and prospective use have little meaning unless related to the situation in particular basins and aquifers. Such study points concretely to the existence of limits of supply and to the importance of carefully appraising the options in making wise use of it. Parallel to and preceding the IHD was the International Biological Program which stimulated in a somewhat similar fashion collaboration among biologists concerned with assessment of productivity of terrestrial fresh water and marine ecosystems, the conservation of terrestrial ecosystems, the use and management of such systems, and human adaptability. This provided important new information on the nature of the aquatic systems perturbed by water development.

Geographic Appraisal of the Results

Some of the efforts have fallen far short of achieving their aims while others have been conspicuously successful in advancing economic growth and social stability. Unanticipated effects have been generated by many projects, particularly as they affected aquatic and terrestrial ecosystems and social organization of the people who had been expected to benefit.

In such hindsight appraisal there has been a persistent tendency on the part of the press and semitechnical writers to paint pictures of distressing side effects of large water projects. The great multipurpose dam which in midcentury was a symbol of social advancement and technological prowess came into bad odor by the 1970s and was often attacked as destructive and poorly conceived. The more substantial effects of the projects in increasing the supply of cheap energy, enlarging irrigated land, reducing transportation cost, and assuring moderately potable urban supply tended to be taken for granted. However, it would be as narrow to report only the successes of large-scale river development as to concentrate upon its failures. Thus far, there has not been sufficiently profound and precise analysis to permit a balanced examination of the full consequences of river development for more than a few areas. To do so would require a series of genuine post-audits which are made only in special circumstances and which at best are difficult to achieve.

Geographical analysis has played a modest part in this unfolding set of approaches to river development, and at present joins in the few systematic efforts to examine what has happened as a result of investment in river development and what lessons may be learned for future ventures in this field. The design, planning, construction, and operation of works for water management usually have been in the hands of engineers, hydrologists, and economists. Some of the work of geographers has been outlined in earlier papers.[3]

Since the publication of those papers there have been a few notable advances in the kind of examination which geographers have made. These include a series of critical studies of river basin development activity. The evolution of the Indus experience has been traced with meticulous care.[4] The effects of irrigation initiatives in Sri Lanka were estimated by Farmer.[5] The lower Rio Grande programs have been shown by Day to satisfy the conditions of amicable relations between Mexico and the United States while failing to achieve physical and economic aims.[6] Examinations of flood problems in selected areas of the world led to the development of new public policies to cope more effectively with floods.[7] A few other post-audits of river basin projects have been made. The problem of social response to reuse of urban waste water has been a matter of recent emphasis by geographers.[8] Probably more than any other academic group they have engaged in critical review of the role of public perceptions and attitudes in affecting the design and operation of water management activities.[9]

The appraisals presented in this volume are an effort to draw upon a wide variety of scientific data and analysis and upon the judgment of an International Geographical Union (IGU) symposium in asking a few rudimentary questions about international experience with the development of selected river basins. They are neither comprehensive in coverage of the continental land surface nor exhaustive in review of the array of troublesome problems presented by water development. They nevertheless provide a comparative examination of experience in a few major basins. From these and from scattered studies of other basins it is possible to draw generalizations as to scientific problems which should be canvassed if future development of rivers around the world is to serve its essential aims of assuring stable supplies of water for domestic use, industrial processes, energy generation, and food production. To assure such supplies requires a form of development which will maintain the basic resources of land and water free from deterioration in capacity to support life.

Critical review of experience with complex utilization of rivers necessarily involves analysis of the spatial distribution of water, land, people, and human organization within the area affected, the technologic and social devices for water and land management, and the identifiable environmental impacts from development programs. No solid appraisal of the programs for the Plate, the Columbia, or the Volga can be realistic without examining shifts in spatial patterns. It needs to ask what measures have been applied among the large number of activities that are possible in complex river development. The full consequences for land, water, and atmosphere also need to be examined so far as practicable.

Selected Scientific Problems

Three major problems which emerge from geographic analysis are: (1) assessment of the possibility and implications of long-term fluctuations in the supply of water; (2) the narrow range of possible adjustments which commonly are canvassed in planning changes in spatial patterns of water; and (3) the prediction of environmental consequences of water development. Each of these has implications for organization of the scientific and administrative activities touching on them.

Long-term Climatic Change

Virtually all major programs for complex utilization of rivers assume that the average conditions for which hydrologic records are available at the time plans are drawn for the project will continue into the indefinite future. To proceed on this assumption leads to difficulties in the long run from having either overestimated the supply of water or underestimated the demand for it by plant growth.

The Colorado basin in North America illustrates the troubles arising from having estimated basic water supply during a wet period, followed by a long dry period in which the planned works were put into operation and in which there was persistent shortage in meeting both demands for irrigation and the allocations of supply which had been arrived at by political negotiations and embodied in interstate compacts and international treaties.[10]

Use of precipitation and stream flow records for forecasting future conditions is attractive to the design engineer. It permits rigorous statistical analyses and extrapolation. Unless synthetic time series are generated by combining the records of similar areas or by stochastic techniques, the historical measurements may seem all that is available. Nevertheless, they can be supplemented in two important ways. First, evidence of previous climatic conditions can be pieced together from tree rings, clay varves, and pollen analysis to indicate whether the period of record is wetter or dryer than earlier times. Second, models or observations of changes in stream flow in relation to upstream landscape modification aid in estimating how much the quantity and quality of water supply may be altered by future land use. The latter form of analysis is an important component of planning for the Volga, and it increasingly receives attention in areas of rapid urban expansion of roofs, streets, and drains.

The current state of climatology does not permit the assignment of causes to what appear to be short-term trends, let alone to long-term changes in the amount of precipitation or in average temperature. There are theoretical grounds for believing that changes in albedo resulting from forest cutting, different patterns of land use, particulate matter in the atmosphere, and the atmosphere's CO^2 content may be related to shifts in climate. This body of evidence and theory was reviewed by Hare in a report for the United Nations Desertification Conference but does not yet permit firm predictions about the course of weather in the decades ahead.[11]

Lacking the theoretical grounds for predicting climatic change, water planners will be well advised to go beyond the meteorological records to the reconstruction of previous environments as a means of checking on the validity of the conventional extrapolations of the recent past. They also will gain from explicit estimates of how the hydrologic conditions may be expected to change under the hand of man.

Broadening the Range of Adjustments

A basic concept which pervades much current planning for water management derives from geographic analysis. It is the view that for any given environmental situation there is a wide range of alternative adjustments among which societies may choose in arriving at a suitable plan of action for a given period of time.[12] While simple in theory this concept is extremely difficult to apply. Once applied, it provokes radical changes in design.

Societies tend to look to the future in the perspective of whatever sets of adjustments to natural resources have survived in the past, and to focus on a few technical improvements, generally promoting new adjustments with which they are familiar or for which they feel a special competence in planning and operation. An irrigator is inclined to go on irrigating his fields by whatever practice has proved effective in the past. At the same time, the irrigation engineer or public works administrator who canvasses improvements in irrigation practice is likely to think of methods which he or she has been trained to apply. For several decades it was common for engineers to plan new irrigation projects without making use of the facilities for a combination of pumping and sprinkler application which were applicable to quite diverse terrain and decentralized management. Over time these techniques have become more widely distributed and adopted, and it now would be expected that a new irrigation project would take them into account in deciding which system of management to employ. In this process the older methods, such as the flood irrigation of the lowlands bordering the Don, may be neglected with resulting threats to marsh and aquatic ecosystems.

The common methods of planning for water management thereby have a major weakness. The plans usually fail to take account of the full range of alternatives that are available for water management—they ignore traditional or peasant methods. All too often the stable adaptations of the past are expected to be superceded by new ones taught in the engineering, agricultural, and scientific institutions of western countries.

The expanded perspective has large implications for national economic plans because it employs a wider range of economic and social tools. These include management of water demand in contrast to management of supply, pricing of electric power, control of effluent disposal by industrial processes, regulation of floodplain use, and the like.

One forceful demonstration of the broadened use of alternatives is in the flood control field. With the spread of both single-purpose and multipurpose river engineering measures heavy reliance was placed on dams, levees, and channel improvements for flood loss reduction, as demonstrated in the Fraser basin.[13] In recent years attention turned to use of nonstructural measures, including the flood-proofing of buildings, use regulations, subdivision regulations, building permits, and land acquisition. Another example is in the field of water pollution control. The alternative of preventing waste from reaching a sewer system in the first place is considered along with modes of treating effluent from residential and industrial sources.

A social device of information, research, taxation, or pricing may be as sound and influential in achieving water development as an engineering measure. The use of economic incentives for guiding the course of water development has gained more attention in the Western European scene than elsewhere and it offers promising opportunities to curb pollution from point sources. In highly intensive systems of management, such as are found in Israel, the use of such devices may point the way to strategies in other nations as they approach physical limits of water supply and as they cultivate the administrative capacity to exercise refined management techniques.

The elementary notion is that there are a large number of alternative adjustments which may be made to environmental conditions. To concentrate upon one measure to the exclusion of others and to fail to seek optimal combinations of measures suited to the local landscape is to court economic loss and landscape degradation.

Finding the optimal measure or mix of measures is a frustrating task. Perhaps the most far-reaching changes in techniques for managing water have come in the methods of appraising social consequences of development and for arriving at judgments as to optimal allocations of capital. The methodology of benefits and costs estimation has reached a highly refined state, but the institutional administrative mechanisms for using such calculations lag behind.[14] Similarly, the elaborate models of surface and groundwater systems which provide a possible basis for economic optimization of water resources development have exceeded the capacity of scientists to specify all pertinent parameters and of political agencies to use them as tools in decisions as to allocation of water or capital funds.

Estimating Environmental Impacts

For each of these general sectors of inquiry two kinds of questions may be asked: (1) What is known about the effects of past development programs on the environment? and (2) What appear to be the principal future effects of alternative adjustments? The preparation of estimates of environmental impact has become an exercise whose significance is widely recognized by national governments and international agencies. The problems raised are like many discussed in the following papers in this volume. To what extent has the fertility of the soil been affected by works for flood protection, irrigation, or drainage? To what extent has the natural channel of the stream been altered in cross section and in gradient with resulting changes in the transport of sediment and in the accumulation of silt in the channels and deltas of those rivers? To what extent has the quality of the water in the river or in the reservoirs been altered in its mineral composition as well as in its bacterial load, including viruses, and the presence of toxic substances which constitute chemical contaminants? Many other questions could be added to the list.

Geographical analysis of water management emphasizes the importance of approaching future management from a historical point of view. It makes it plain that management evolves from one period to another, and that in order to understand the impacts of future action it is essential that the distinctive requirements and opportunities for solving past problems be recognized. A conspicuous aspect of water management has been the lack of careful post-audits of the social, economic, and environmental consequences of previous works. The making of a post-audit is not an easy task. It can only be as complete as the number of impacts that are examined. It can only be as precise as the scientific method available to trace out those impacts and to assign values to them. The number of careful efforts to examine the full range of consequences have been relatively small. Among the more careful post-audits from the geographic literature are the studies of the Uncompahgre irrigation project, the Sabi Valley irrigation projects, the monumental volumes by Michel on the Indus and the Helmand, and Day's study of the lower Rio Grande.[15]

During the past two decades concern for recognition and measurement of the character of the environmental impacts has mounted rapidly. An example is in the case of examination of effects of storage impoundments behind large dams, particularly in tropical areas. The past fifteen years saw the initiation of a number of investigations on such projects, including Kariba, Volta, Kainji, high Aswan, and lower Mekong projects, to assess the actual and potential consequences of the new programs. Impacts included the change in the aquatic ecosystem with alteration in fish population and production, the use of adjoining terrestrial ecosystems, possible cultivation of seasonally inundated reservoir margins, social consequences of population relocation, the hazards to health resulting from schistosomiasis, malaria, and the like, and water transportation. These investigations were financed by the United Nations Development Programme, and executed by the Food and Agriculture Organization (FAO) and the World Health Organization (WHO).[16]

Most of these studies have been in the nature of salvage operations. They seek to pick up the pieces of ill-considered and short-sighted administration after the projects are underway. To a lesser degree similar investigations have been initiated on new projects before they reach the construction stage. As a result of policies adopted by the major funding agencies at the international level, the likelihood of any such works going into construction without such preliminary investigation has been greatly reduced but not eliminated.

Scientific speculation is just beginning to direct adequate attention to the question of how much the construction of new reservoirs, channel diversions, and waste discharges will affect the climate of large areas and the chemical composition of ocean water. The possibility of diverting north-flowing streams in North America and Eurasia to the south raises the issue to a level of intensive investigation having international implications. The dimensions of the question are illustrated in the current examination of projected interbasin water transfers in the USSR.

Lessons from the Experience

It is apparent from the geographic appraisals presented in the Volga-Don symposium that scientists in different countries concerned with water management experience are convinced that heightened attention needs to be given to anticipating problems of maintaining environmental quality. Specifically, they emphasize a historical perspective, greater precision in forecasting physiogeographical effects and stream flow, a more diverse strategy in managing water quality, and maintaining the integrity of river systems affected by interbasin transfers.

Major improvement in future management could be gained by approaching water planning from a historical point of view. This means examining projected water management in conjunction with past modes of managing land, mineral, and related resources. In each natural area it is possible to recognize distinct historical stages of economic development and resource management. Each had its own requirements for solving problems of water and environmental conditions. Each region of the continents presents different issues and calls for different combinations of solutions. But planning for the next steps should be taken in recognition, as in the case of the Volga or the Tana, of conditions leading to previous stages and the constraints and opportunities which they presented.

In looking to the future, the making of physiogeographical forecasts of the effects of major water projects becomes increasingly important as the requisites of economic development change rapidly and as the available technologies for use and misuse of water resources multiply. Basic to all water planning are forecasts of the seasonal quantity and quality of water flow. Modern research techniques tend to concentrate on mathematical, quantitative models. These continue to be refined by more sophisticated modes of data processing and statistical analysis, including the simulation of long time series. These methods need to be combined with various forms of geophysical, biological, and physical analyses which take account of the evidence available from analysis of tree rings, varves, soil profiles, and sedimentary deposits in reconstructing past climates and the sequence of hydrologic events preceding the organization of modern meteorological observation networks. With much less accuracy than meteorological records, it is possible to estimate long-term environmental fluctuations so that forecasts of future stream flow will be leavened by evidence of swings in precipitation and temperature, whether or not those involve secular trends in climate. Thereby, the trap of planning for a Colorado River flow on the basis of records unlikely to be duplicated in the next century would be avoided.

The medium- and long-range forecasts of stream flow should take account of the prospective effects of changes in watershed conditions as they may alter evaporation, transpiration, soil storage, and stream flow. The dramatic cases of rapid change are in the urbanization of small sections of watershed. In the long run, changes in agricultural and industrial land use may be of large consequence. Explicit provision should be made in forecasts for such changes.

Efforts of this sort should not be made on a narrow, regional basis. Experience in one part of the world may be relevant to another even though the particular combinations of landscape features may differ. It is therefore desirable to accelerate international exchange of information on changes of precipitation, groundwater, and streamflow as inferred from physical and biological sources.

A marked change in emphasis should take place in managing water quality. Earlier emphasis upon quantitative control of flows and storage facilities needs to be supplemented by more varied and sophisticated methods of dealing with quality. These would involve at least three directions of change. First, the treatment of waste water, including water reuse and the substitution of waterless or wasteless technology, deserves increasing attention. If systems like the Plate, the Columbia, and the Volga are to be maintained in stable quality it will become essential to drastically reduce the volume of waste discharge into these water bodies and to explore every practicable means of reducing waste. It is not unrealistic to think of a time when there will be no waste in the traditional sense but only full use of all materials that enter into life-supporting systems.

Second, to the extent that there is discharge of waste into waterways it is important to optimize the sites for discharge in relation to the absorptive capacity of the stream system. In some situations it may be essential to bury non-utilized waste products, but the long-term goal would be to eliminate the need for such palliatives.

Third, emphasis should be placed on maintaining and intensifying the self-purification capacities of soils, fresh waters, and sea waters, as illustrated in the Azov Sea. To the extent that products of land use contribute to the nourishment of aquatic ecosystems, their addition should be encouraged, but it needs to be finely tuned to the capacities of the systems to utilize the materials and to produce useful products in the form of fish catch.

One of the observations that arises most acutely from the Volga experience is the necessity of artificially reinstating the conditions of periodic flooding which characterize virtually all natural stream systems. Earlier water management tended to control floods, storing the peak discharges in reservoirs or diverting them into other stream systems or basins. It is now recognized that it may be essential, for the sake of preserving stream capacity to utilize waste and maintain aquatic organisms, to provide for periodic flooding by controlled releases from reservoirs. While avoiding the great peaks that would cause catastrophic damage, it will become essential in many streams to regenerate flood flows as a means of maintaining natural environmental processes.

Increasingly, geographers find it desirable to look to the integrity of natural systems, such as river–sea, river valley–flood plain, and river channel–catchment area relationships, in order to maintain natural processes.[17] This may heighten conflicts between different branches of national economy. There is a continuing strain between power development and fisheries, between irrigation and fisheries, and between different territories within a river basin as illustrated in the conflicts over water in the Volga, the Fraser, the Plate, and the Columbia. These conflicts cannot be fully eliminated but in many instances can be softened. The important aim is to arrive at compromises which still take account of the natural water and land processes.

The experience in countries other than the USSR and the United States has recently begun to be examined in a critical fashion. There is much to be learned from the successes and failures in the Danube, the Rhine, and the streams in Canada, India, and the People's Republic of China, which have been developed extensively in recent years. It would be useful to put together a series of long-term post-audits of national efforts in basins such as the Danube where more than one country is involved.

Special attention should be called to the possible consequences of major interbasin transfers of river runoff. These problems should occupy a special place in future geographic research. Before momentous and possibly irreversible decisions are made with respect to those projects there should be a more nearly incisive examination of the environmental consequences. The needed research should consider the interests of consumers and donors of water and the effects of transfers on the southern and northern seas. In thinking of interbasin transfers in North America and in the USSR (such as those being considered for the transfer of north-flowing streams), the utmost care should be given to the potential climatic and other effects of these new large-scale interventions.

Fundamental research on transfers of river runoff should not prohibit consideration of opportunities and problems associated with small local transfers to meet urgent contemporary needs. However, the proviso should always apply that such measures also should be assessed in terms of their effects on the maintenance of the natural river and land processes.

As illustrated by the reports in this volume, different countries plan in different ways to solve problems unique to their environment. It is not practicable to specify what optimal water management would comprise in every area. Yet, the scarcity of water resources in many areas and the continuing degradation of water quality strongly suggest caution in interventions where the full array of environmental consequences have not been investigated.

Developing countries present special problems of water management. They should profit from the lessons learned in the nations which have made heavier investments in engineering works. They are under urgent pressure to use water management to accelerate economic and social development. Nevertheless, their capital resources are small, and they risk permanent harm to natural processes in attempting to generate short-term plans for economic growth. The economic demands need to be reconciled in some logical and sensitive fashion with consideration of long-term capacity of the environmental system to support population.

Rivers crossing international frontiers are a special case for investigation. As indicated by the experience with the Columbia and the Plate, the difficulties of generating cooperation among riverine countries are substantial, but not wholly insoluble. Much can be learned about what to avoid from the efforts in the Danube, Columbia, and Arax. It is possible for nations to meet each other at some roughly halfway point in order to satisfy the needs of one generation without jeopardizing the interests of future ones. Severe international questions loom for rivers in Latin America, South Asia, and Africa. They deserve more systematic attention both in terms of understanding what has happened in the developed streams and what special conditions apply to the initiation of new ventures.

It is plain in the reports, presented in the following chapters, of past efforts at complex river development in the Volga, Don, Columbia, Fraser, Plate, and Tana basins that there is no uniform approach to assessing environmental effects. Certain areas are far advanced in identifying and evaluating the actual or likely consequences of human intervention in water and land systems. Other areas have only begun to incorporate such assessment in development planning. As pointed out in the

Fraser Basin study, the policy and administrative structures have been slow to respond to recognized difficulties in managing environmental impacts, and often the change is stimulated by biological or social stress. However, the trend in these and many other basins is toward more accurate definition of effects and of correlative research requirements.

When international experience with complex river development in sectors of Africa, Eurasia, North America, and South America is examined in the light of historical trends, the mix of human adjustments practiced, and the probable environmental impacts, a few needs stand out. Long-term climatic and landscape changes need to be assessed more carefully. The tactics of manipulating water quality and quantity need to be appraised more fully with a view to employing a broader range, including social measures, of alternatives. Critical post-audits of what in fact has happened in the trail of large-scale development should be expanded. The great rivers of the world provide striking, earthy lessons that can be ignored only at the cost of environmental degradation.

Notes

1. United Nations, *Integrated River Basin Development: Report by a Panel of Experts* (New York: United Nations Department of Economic and Social Affairs, 1958, revised 1970).

2. M. Lvovich, *Global Water Resources and Their Future* (Moscow, 1974); and Hilgard Sternberg, "The Amazon River of Brazil," *Geographische Zeitschrift* (Beihefte, Erdkundliches Wissen, Heft 40, Wiesbaden, 1975).

3. Richard J. Chorley (ed.), *Water, Earth, and Man* (London: Methuen, 1969); Francis M. Leversedge (ed.), *Priorities in Water Management*, Western Geographical Series, Vol. 8 (Victoria, British Columbia: University of Victoria, 1974); Gilbert F. White, "Geographic Contributions to River Basin Development," *Geographic Journal,* Vol. 129 (1963), pp. 412-436; and Gilbert F. White, "Role of Geography in Water Resources Management," in *Man and Water: The Social*

Sciences in Management of Water Resources, edited by L. Douglas James (Lexington: The University Press of Kentucky, 1974), pp. 102-121.

4. Aloys A. Michel, *The Indus Rivers: A Study of the Effects of Partition* (New Haven, Conn.: Yale University Press, 1967).

5. B. H. Farmer, *Pioneer Colonization in Ceylon: A Study in Asian Agrarian Problems* (London: Oxford University Press, 1957).

6. John C. Day, *Managing the Lower Rio Grande: An Experience in International River Development* (Chicago: University of Chicago, Department of Geography, 1970).

7. United Nations, *Report of the United Nations Interregional Seminar on Flood Damage Prevention Measures and Management,* ST/TAO/SER.C/144 (Tbilisi, USSR, Sept. 25-Oct. 15, 1969); and President's Task Force on Federal Flood Control Policy, *A Unified National Program for Managing Flood Losses* (Washington, D.C.: Government Printing Office, 1966).

8. Leonard Zobler et al., *Benefits from Integrated Water Management in Urban Areas* (New York: Barnard College, 1969).

9. Duane Baumann, *The Recreational Use of Domestic Water Supply Reservoirs: Perception and Choice* (Chicago: University of Chicago, Department of Geography, 1969); Robert W. Kates, "Variation in Flood Hazard Perception: Implications for Rational Flood-Plain Use," in *Special Organization of Land Uses: The Willamette Valley,* edited by J. G. Jensen (Corvallis: The University of Oregon, 1964); A. H. Laycock, "American Attitudes Concerning Canadian Water," *Albertan Geographer* 7 (1971), pp. 24-33; Thomas F. Saarinen, *Perception of the Drought Hazard on the Great Plains* (Chicago: University of Chicago, Department of Geography, 1966); and Derrick Sewell, Richard W. Judy, and Lionel Qvellet, *Water Management Research: Social Science Priorities* (Ottawa: Department of Energy, Mines, and Resources, 1969).

10. National Academy of Sciences, Committee on Water, *Water and Choice in the Colorado Basin: An Example of Alternatives in Water Management* (Washington, D. C.: Na-

tional Academy of Sciences, 1968).

11. F. Kenneth Hare, *Climate and Desertification,* Preliminary draft report prepared for the United Nations Environment Program (University of Toronto: Institute for Environmental Studies, 1977).

12. Ronald R. Boyce (ed.), *Regional Development and the Wabash Basin* (Urbana: University of Illinois Press, 1964); Robert K. Davis, *The Range of Choice in Water Management* (Baltimore: Johns Hopkins Press, 1968); National Academy of Sciences, Committee on Water, *Alternatives on Water Management* (Washington, D.C.: National Academy of Sciences, 1966); Donald J. Volk, *The Meramac Basin: Water and Economic Development* (St. Louis: Meramac Basin Research Project, 1962); and Gilbert F. White, *Strategies of American Water Management* (Ann Arbor: University of Michigan Press, 1969).

13. Robert W. Kates, *Hazard and Choice Perception in Flood Plain Management* (Chicago: University of Chicago, Department of Geography, 1962); United Nations, *Report of the United Nations Interregional Seminar*; and President's Task Force on Federal Flood Control Policy, *A Unified National Program.*

14. Marion E. Marts and W. R. Derrick Sewell, "The Application of Benefit-Cost Analysis to Fish Preservation Expenditures: A Neglected Aspect of River Basin Investment Decision," *Land Economics,* Vol. 35, No. 1 (February 1959), pp. 48-55.

15. Jacquelyn L. Beyer, *Integration of Grazing and Crop Agriculture: Resources Management Problems in the Uncompahgre Valley Irrigation Project* (Chicago: University of Chicago, Department of Geography, 1957); Aloys A. Michel, *The Kabul, Kunduz, and Helmand Valleys and the National Economy of Afghanistan* (Washington, D.C.: National Academy of Sciences–National Research Council, 1959); Wolf Roder, *The Sabi Valley Irrigation Projects* (Chicago: University of Chicago, Department of Geography, 1965); Clifford Russell, David Arey, and Robert W. Kates, *Drought and Water Supply* (Baltimore: Johns Hopkins Press, 1970); S. L. Vendrov, "Water Management Problems of

Western Siberia," *Soviet Geography: Review and Translation,* Vol. 5, No. 5 (1964), pp. 13-24; and M. Gordon Wolman, "Two Problems Involving River Channel Changes and Background Observations," in *Quantitative Geography: Part II,* edited by W. L. Garrison and D. F. Marble (Evanston, Ill.: Northwestern University, 1967).

16. William Ackermann, Gilbert F. White, and E. B. Worthington (eds.), *Manmade Lakes: Their Problems and Environmental Effects,* Geophysical Monograph No. 17 (Washington, D.C.: American Geophysical Union, 1973).

17. A. Ryabchikov, *The Changing Face of the Earth: The Structure and Dynamics of the Geosphere, Its Natural Development and the Changes Caused by Man* (Moscow: Progress Publishers, 1975); A. N. Voznesenskiy, G. G. Gangardt, and I. A. Gerardi, "Principal Trends and Prospects of the Use of Water Resources in the USSR," *Soviet Geography: Review and Translation,* Vol. 17, No. 5 (1975), pp. 291-302; and Gilbert F. White et al., "Economic and Social Aspects of Lower Mekong Development," A report to the Committee for Co-ordination of Investigation of the Lower Mekong Basin, 1962.

2
The Volga River

S. L. Vendrov and A. B. Avakyan

The Volga is the embodiment of all Russian rivers. Is is not accidental that since ancient times the Russian people have tenderly called it the "Dear Mother Volga." At all stages in the nation's history the Volga has been one of the main axes of population settlement and of the development of the country's economy and culture.

Outstanding sociologists, politicians, historians, and economists, eminent figures in literature and arts, mathematics, chemistry, biology, and other sciences were born, grew up, received their education, and worked in the towns and cities on the Volga banks. During World War II the industry and agriculture of the Povolzhiye served as an important material and strategical base for resistance against the enemy's intrusion. The Stalingrad battle was the turning point in the defeat of German fascism and the victory gained by the Soviet Union and its allies.

S. L. Vendrov is professor, Institute of Geography, Soviet Academy of Sciences, Moscow. A. B. Avakyan is professor, Institute of Water Problems, Soviet Academy of Sciences, Moscow. Abstracts of this chapter were published in "Man and Environment," *Symposium of the Commission: Abstracts of Papers and Presentations* (in Russian) (Moscow: Vneshtorgizdat, 1976).

The Basin and Its Use

The Volga is the largest river in Europe.[1] It is 3,350 km long and its major tributaries, the Kama and the Oka, are 1,800 km and 1,500 km long respectively. The length of all the Volga basin rivers over 10 km long (see Figure 2-1) is 270,000 km, and the total length of the entire basin river network comprises 572,000 km. The total drainage area is 1,360,000 km[2], embracing the forest, forest-steppe, steppe, semidesert, and desert geographical zones. The basin occupies the center, east, partially northeast, and southeast regions of the European USSR (within the boundaries of the RSF).

The annual Volga runoff is 243 km[3], constituting 80 percent of the average inflow to the Caspian Sea.[2] The Volga, before its confluence with the Kama, gives 119 km[3] of this volume (the Oka gives 32 km[3]), and the Kama brings 124 km[3]. In years with low water capacity (95 percent) the Volga's runoff is only 173 km[3], and that of the Kama is 81 km[3].

The ratio between the low water capacity runoff and the average runoff on the lower Volga is 0.73; on the section above the confluence with the Kama it is 0.68; and on the Kama it is 0.63. The eastern (Kama basin) and western parts of the Volga drainage area differ in their environmental conditions, and often in their synoptics. Therefore, the intra-annual variations of the upper Volga and Kama runoffs are not in phase, making the annual fluctuations of the lower Volga runoff somewhat more uniform as compared with the other rivers of the European USSR.[3] The intra-annual variations in the Volga's natural runoff can be illustrated by the following comparison: the snowmelt floods of April, May, and June constitute from 50 to 60 percent of the annual volume while the other nine months constitute only 40 to 50 percent of it.[4]

Hydrologically, biologically, and economically the Volga basin is closely related to the Caspian Sea. They are traditionally regarded as one common Volga-Caspian problem, the essence of which is as follows: the stage level of the Caspian Sea has no outflow and consequently, its nature and ecology (in particular, of the northern Caspian which is an extremely rich fishery region) largely depends upon the Volga water, solid

FIG. 2–1
THE VOLGA RIVER

matter, and ionic and biogenic discharge, which varies greatly due to climatic fluctuations (intrasecular and of longer periods) and anthropogenic effects.

The Volga-Caspian biological resources have been widely used at many historical stages of economic development. At present the internal reservoirs of this basin are responsible for 50 percent of the total USSR fish catch and 90 percent of the sturgeon catch.

The use of the water and biological resources of the Volga basin, especially of the Volga and its major tributaries (the Kama and Oka), has always been complex. The nature and degree of this complex usage is changing along with the leading branches of the national economy which are the primary consumers of these resources. Each basin has one or several of such "leading branches" which specify the capital investments and trends in the development of the entire complex. For the Volga these leading branches have combined in the following sequence: (1) until the thirties of this century transport and fishery went together; (2) hydropower generation and water transport, with water supply and thermal electric engineering increased in significance during the past 10–13 years; (3) paralleling this change, agriculture, and irrigation in particular, became more and more important; (4) water supply to the adjacent basins of the Don river and the Azov sea, of the Ural river and the Caucasus recently became an urgent task for immediate solution; and (5) the programs and current plans for using the Volga are devoting more attention to the problems of recreation.

At present, in discussing the regime and use of Volga runoff, the interests of the Caspian Sea and the Azov Sea and its basin are always taken into account. Current views of water economy regard these two seas as an important water consumer of flows from the Don and Volga systems. Plans for a more rational inflow of fresh waters into these seas should concentrate on further promotion of the fisheries in the internal reservoirs.

As on all intensively used rivers, the water quality problem has become acute on the Volga. The question of pollution control on this river has always been a focus of attention. Thanks to the measures taken since 1972, water pollution on the Volga and its tributaries has considerably decreased. On some sections it has practically stopped. The measures to protect the Volga waters from pollution include the following: (1) further improvement of treatment techniques, both by special treatment plants and by nature itself, introduction of closed water supply systems and reuse of secondary wastes, elaboration of new wasteless (and in some cases, waterless) technologies; (2) optimization and rationalization of the waste discharge sites and, if necessary, burial of the nonutilized remains of the waste waters; and (3) further preservation and intensification of the self-purifying capacity of the rivers, lakes, reservoirs, and estuaries. This problem is related to limited environmental loads. On the regulated rivers, like the Volga and some of its tributaries, and also on all water economy systems, it is necessary to generate special periodical washoffs which act as natural floods. Such washoffs greatly improve the local sanitary and ecological conditions.

Hydroengineering construction on the Volga was initiated in the late thirties and then, after a halt caused by World War II, it was renewed and intensified in the fifties and sixties. Nine hydroelectric stations have been built on the Volga and the Kama. Two stations are now under construction. Their total fixed capacity is 12 million kw, and the annual average output is about 40 billion kw/hr. The two largest stations, one upstream from the city of Kuibyshev (the V. I. Lenin hydrostation) and the second one upstream from the city of Volgograd (the 22nd Party Congress Station), each produce 10–11 billion kw/hr annually. The Volga and Kama hydroelectric stations are mighty strongholds of the integral energy system of the European USSR. They cover all peak loads and function as frequency and emergency standbys. With this advanced hydroengineering construction completed, the Volga and Kama have turned into rivers with cascades of reservoirs (see Table 2-1).

These reservoirs regulate the Volga and Kama seasonally. Reservoirs with multiannual regulation are not possible without creating excessive flooding. All in all, there are more than 200 reservoirs in the Volga basin, including the small ones. Their total area is about 24,000 km², and their effective volume, which makes up 37 percent of the average annual runoff, is 90 km³. The Volga runoff losses caused by evaporation from the reservoir surfaces are approximately 7–8 km³. This estimate takes into account that the evaporation from the reservoirs in the northern part of the basin is smaller than the transpiration losses typical for these areas before they were submerged.

All reservoirs on the Volga and Kama are multipurpose and used in the following ways: for hydropower, thermal power, water transport, industrial and community water supply, irrigation (except the reservoirs in the northern part of the basin), fishery, and recreation.

TABLE 2-1

THE VOLGA AND KAMA RESERVOIRS

N,N	Name of the Reservoir	Water-surface area at normal back-water level/km²	Reservoir volume/km³	
			Total	Effective
1)	Ivankov	327	1.12	0.81
2)	Uglich	249	1.25	0.81
3)	Rybinsk	4550	25.42	16.67
4)	Gorky	1590	8.82	3.97
5)	Cheboksary (underway)	2190	13.85	5.7
6)	Kuibyshev	6450	57.99	34.65
7)	Saratov	1830	12.87	1.75
8)	Volgograd	3120	31.45	8.25
9)	Perm (Kama)	1915	12.2	9.83
10)	Votkinsk (Kama)	1065	9.36	4.45
11)	Lower Kama (underway)	2580	12.9	4.4

The reconstructed Volga is part of the integral transport system of deep-water ways in the European USSR. Its present day cargo movement, which constitutes 75 percent of the cargo on all the domestic waterways of the USSR, is 150 billion t/km. The tonnage of modern self-propelled freight boats is from 2,000–5,000 tons, and in tows it reaches 6,000–15,000 tons. The advantages of a modern river fleet have made it possible to provide for mixed transports (river-sea), for example, from the Caspian or Black seas to the Baltic. The main cargoes carried on the Volga and Kama are timber, grain, minerals, construction materials, cars and other machines, and a variety of raw materials and containers. The passenger fleet was greatly expanded in recent years.

The Volga waters are widely used for irrigation. At present they supply more than 500,000 ha. Irrigation with Volga waters has almost eliminated the effects of droughts which are so frequent in the regions of Transvolga, the Povolzhiye, and the eastern part of the front Caucasus.

Economic use of the Volga, especially within its basin, has changed and is still changing the nature and character of the bilateral interrelations of the river and the natural environment of its drainage area.

Environmental Effects

The improvement of navigation conditions on the Volga, Kama, Oka, and lesser rivers which has been underway since ancient times has intensified during the last twenty-five years.[5] Special projects to clean the banks and deepen the bottoms of these rivers were undertaken, particularly on shoals which are so numerous on the Russian plain rivers in their natural flow. When these modifications are undertaken on a large scale and cover long sections of transit routes they actively alter the environment by affecting the relationship between water level and discharge and the channel depth in upstream reaches and at the river's mouth. This reforms the river bed and changes the stream hydraulics. The ecology of the river basin is changed accordingly. In their scale and scope, these changes cannot compare with the transformation of the river system into a manmade cascade of reservoirs, and the tributary territories into subordinate aquatorias. It is important to emphasize that the problem of environmental changes and conservation while using the environment has become extremely acute in recent times. It is, however, a very old problem, as, indeed, are many of the other so-called new water problems. It acquired fresh meaning in connection with the greatly increased loads on the environment caused by advances in modern economy and culture.

The effects of interbasin linkages of water transport in past centuries upon the watersheds and the environment were only local. Nevertheless, their role in the economy of the basin was as important as that of the later bottom-deepening measures. They provided links between the rivers of the Caspian and

Baltic and the Black and the Baltic seas for the small boats of that time with low tonnage.

The reconstruction of the Volga was initiated in the thirties when the Ivankov, the Rybinsk (filled with water during the war), the Uglich reservoirs were built. During the same time period many hydroelectric and transport facilities were constructed. It was in those years that the waters of the Volga were connected with the middle Moskva River. All of these measures were necessary to secure the water resources needed for the present-day progress and advance of the USSR capitol. In the fifties and sixties the reconstruction of the Volga and its tributaries was further intensified. A cascade of large hydroelectric stations and reservoirs was built on the Volga and Kama. The Volga became an important supplier of power to the integrated energy system of the country, its main contribution being in peak energy.

In the early fifties a canal between the Volga and the Don was completed, thus providing a water connection from the Volga basin (including the center of the country) into the Azov and Black seas. In the sixties, the construction of the Volga-Baltic canal was completed. The new canal replaced the old Mariin system by opening a deep-water channel into the basin of the Neva River and the Baltic Sea.

In 1976 the next step in Volga reconstruction is about to take place. The great river will be connected by interbasin links with the rivers of the European USSR. The Volga will then carry their waters into the northern seas, and, in turn, a comparatively small proportion of those waters will be directed back into the Volga basin.

The new water diversion system will have substantial impacts on the ecological conditions of the Volga. In its natural flow, the stage regime in the Volga, as in any large river, reflects the integrated environmental conditions of different and very diverse parts of its drainage area. The integrated environmental conditions include stream flow, the seasonal changes in flow rate and stage level (including bed), the chemical composition of its waters, its solid matter and biogenic runoff, the specific composition of aquatic organisms living in its waters, and the general ecological conditions. There were, certainly, some local effects on the environment, but they were subordinate, and affected only some elements of the river regime. For example, the time that the ice breaks, the formation of ice, and water temperature, which always varied greatly along the river, are due chiefly to its meridional flow.

The principal impacts from the construction of a large reservoir, and a cascade of reservoirs, are upon the environmental factor interacting with the river. The leading role in the entire process of drainage to the sea is still played by the total stream flow of the river. In all other aspects of the river's regime the commanding role is now played by local factors. Anthropogenic factors influence the stage fluctuations and variations in the flow of the water which is discharged through dams. Natural and anthropogenic factors influence the chemical and biological processes, the solid matter in the runoff, and the formation of bottom sediments and the banks. The character and area of the reservoir's impact upon the land also change. Its waters in the reservoir or in the form of seepage influence the territory adjacent to it and alter local climate, soils, and vegetation. Different types of denudation develop and thereby change the local fauna, including the microorganisms. With the cascade of Volga reservoirs completed, the waters from the upper reaches of the river now run to the delta six to ten times faster. The mixing rate has considerably decreased and it has radically, though gradually, changed the water's natural ability for self-purification.

Spawning travel of migratory and semimigratory species of fish has been hindered drastically. This dictates the replacement of natural fisheries by a controlled and planned fish industry, in particular, by fish hatcheries and spawning farms. Some success in those practices has been achieved in recent years.

Recent observations of the aquatic ecosystems have evoked great interest among scientists. Migrants from the Azov–Black Sea basin have penetrated into the waters of the Volga cascade through the Volga-Don canal. Migrants from the Baltic Sea and its drainage area rivers have penetrated the Volga cascade through the new Volga–Baltic Sea canal. Today, Caspian and Volga species are sometimes found in the waters of the adjacent basins. This is true of some individual species of benthos and plankton as well as of zooplankton. From this evidence it is concluded that migration is achieved not only by individual organisms sticking to the boat bottoms but also in the water flow during the locking process.

The retarded water mixing and the extremely rapid industrial development along the Volga banks and in parts of its drainage area have largely aggravated the pollution of the river waters. As mentioned above, thanks to the corrective measures taken during recent years no enlarged contamination has been observed on the Volga. In fact, its waters are getting cleaner with every new year of its use.

In discussing the changing water quality of the Volga it is proper to clarify the classical definition of a river, as worded by A. I. Voeikov, an outstanding Russian geographer of the turn of this century. He asserted that a river was the product of its own drainage area environment. Today, anthropogenic influence on the drainage area surface affects its arable lands. These lands receive ever-increasing amounts of manmade fertilizers imported from territories far beyond the affected drainage area (for instance, apatites and nephelines from the Kolsky peninsula). The uncontrolled runoff from the fields and forests treated by modern chemicals has provoked essential changes in the hydrochemistry and hydrobiology of small rivers. Increased amounts of nitrogen and phosphorus and their compounds are found in these tributaries. The same applies to herbicides and the like. The waters of the Volga are no longer a product of the natural and anthropogenic conditions of its own drainage area alone.

Differences in Effects

The type of interaction between the two links in the chain—the transformed Volga and its own and adjoining basins' natural environments—varies at different latitudes in accordance with geographical zones. The northern reservoirs of the Volga, up to the Rybinsk, are situated in a zone of excessive humidity. Those northern reservoirs modify the local climate and all components of the landscape in the same way as do natural factors. By adding to the natural factors the continentality of the affected area is decreased. The territory under their influence is equal to or even exceeds the water surface area, and the domain of their episodic influence is much larger. Due to their impact the day temperature in the first half of summer is considerably below the previous state and the summer-autumn transition is also affected. The cloudiness above the huge reservoirs is altered, as is the amount of precipitation on the leeward and windward banks. The agroclimatic parameters, the phenological periods, and the rates of biomass growth all undergo change.

The reservoirs in the Volga's south, namely the Volgograd and the Saratov and the Tsimlyansk Reservoir on the Don River, influence their surrounding territories in a direction opposite to natural climatic and landscape conditions. Their influence fades off much faster, and affects only the territory within a few hundred meters (less than a kilometer as a rule). Their effects therefore spread over an area equal to some 10–15 percent of the water surface area.

The impact of a reservoir upon the aquatic environment differs depending on its position and role in the cascade.[6] The Rybinsk and Kuibyshev reservoirs (similar to the Perm Reservoir on the Kama and the Tsimlyansk Reservoir on the Don), have the largest effective capacities and the highest level amplitudes when compared with the others. Their water surface fluctuations, as well as their periodically exposed bottom areas, are the most extensive. Their environmental effects are more dynamic, resulting in great changes in the natural conditions of the water masses within their own boundaries and far down the river.

The capacity of the Kuibyshev storage basin makes it possible to direct artificial floods from the Volgograd Reservoir into the lower Volga and its delta. This practice is of great economic and environmental importance. Manmade floods depend upon water abundance or deficiency each year, and the discharged water volumes may vary from 60 to 140 km^3 comprising one- to two-thirds of the natural spring flood. They supply water to the Volga-Akhtyubinskaya floodplain and the

Volga delta, thus covering an area more than 500 km long and satisfying the needs of agriculture and fishery. The volume and shape of such discharges are annually determined by the Russian Federation government because any removal of water inevitably results in some economic damage. This damage occurs because the peak discharges of water required cannot be passed wholly through the Volgograd station turbines and therefore are partially sent through the weirs of the dam. The interests of power production and, in part, of water transport can be subordinated to other purposes. When managing water and biological resources in a complex way it sometimes becomes necessary to limit the environmental loads. It is also necessary to preserve some elements of a large river's natural regime in its lower reaches, and in that part of the sea adjacent to the mouth (in this case, it is the northern Caspian). This example deserves special emphasis because flow control along a water artery should always have some limits. For the sake of sanitary factors and the protection of the natural environment and, consequently, for the benefit of the national economy, the regulated waters must provide elements of flood flushing. Such flush elements are indispensable to maintaining water quality in the lower reaches of the river, their deltas and estuaries, and on the coastal zones of the seas.

Because of agricultural and industrial use, a part of the Volga runoff is irreversibly and increasingly removed. This drastically affects the Caspian Sea. Further, it should be borne in mind that the Volga system is vital for additional water of adjacent basins. Those basins include the Ural River to the east and the Don River, together with the Azov Sea, to the west. In view of these demands it becomes urgently important to supply the Volga with waters from the northern rivers which are situated in a zone of excess water supply.

Problems of Transfers from Northern Rivers

Looking to a more distant prospect, say, of fifty years, this desirable supplement of water flow can assume huge values. It may exceed 60 km³ per year on the average. How to achieve such massive transfers of water from the donor rivers is not yet quite clear. It becomes increasingly difficult to solve this problem because of society's constant concern with environmental control. Further fundamental research in environmental processes will be necessary if this important problem is to be solved in the best possible way. Today, scientists are widely discussing serious and practical questions pertaining to the first stage of such water transfers. The discussions concern the most desirable methods to be applied in this century. The scale of the earliest transfer (in several stages) is thought to be of the order of 10–15 percent of the average annual runoff of the Volga, or about 25–35 km^3. That additional volume of water could be very beneficial to the waterways of the southeastern USSR. At the same time, if certain conditions are met, such a transfer would have no negative impacts on the northern territories. These conditions are described below.

Interbasin transfer of river runoff, even in small volumes, should be regarded as more than a measure to improve the water supply of south and southeastern European USSR and the southern seas. It presents an important regional problem as well, involving the interests of the northern territories and seas. Rather than build large reservoirs in the plains territories with the accompanying effects upon land, forest, and other natural resources, it would be more rational to make a wider use of delivery routes which have become more available because of the rapidly growing power output. All kinds of measures to prevent water losses during water transmission (mainly through seepage) should be undertaken. This would avoid unnecessary watering of the land along the new water links in the zone of excessive supply.

The proposed projects call for feeding the Volga through two of its main components mentioned above (the upper Volga with the Oka and Kama) by means of two diversions. When completed, these projects would make it possible to use rationally the irregular fluctuations in waterflow observed for many years in the basins of the donor rivers. Another benefit of augmenting stream flow is that it would help to wash the Volga and its tributary channels. This practice is important for preserving the self-purification capacity of water flux. It should be noted that removal from each donor river must never exceed 10-15 percent of its average annual runoff. Still more important is that it not be done at the cost of streamflow during the river's low-water years.

After some part of the river runoff is removed, sufficient volumes of water must always remain below the hydroelectric station so that the lower reaches, deltas, and estuaries of the northern rivers, which need this cleansing flow as badly as do the southern rivers, can be flooded. Spring floods on the northern rivers begin much later than on the Volga in its lower and middle reaches. Therefore, if the diversion from a northern river takes place when its forecasted flood is past and the actual volumes of the flood have been verified, such removal will prolong the falling time for the Volga reservoirs (their period of fall is shorter when they are filled with the waters of the Volga alone).

Measures should be provided to insure that the quality of the water diverted to the south is not lessened. Those measures would cure anthropogenic pollution of the water and preserve its ecological compatibility for its own and foreign flows. This is one of the highly important factors in the task of inter-basin water transfer into the Volga, even at the initial stages of this enormous and complicated program.

Research Needs

In concluding, attention turns to future research needs generated by resource use on the Volga. At least three types of investigations are important for geographers.

1. It is necessary to study methods of preserving and improving water quality, taking into account the ever-increasing anthropogenic interventions in the drainage area, to help increase bioproductivity of the Volga basin. It is logical, in this respect, to divide the reservoirs into special zones. This would allow their areas, including the shoals, to be optimally improved and most rationally used for fishing, agriculture, recreation, and other purposes.

2. With further study, it should be practicable to optimize

the actual exploitation regimes of the regulated reservoirs in their upper reaches and discharge points. Detailed recommendations on the best possible regimes for meeting the requirements and interests of various branches of the national economy can then be made. It should be kept in mind that sometimes these recommendations will contradict one another and will need to be reconciled. In any case, the investigations should always be sensitive to considerations of environmental protection.

3. Provision should be made for fundamental research and complex geographical forecasts on the possible changes in the environment if more water is diverted into the Volga from the northern rivers than anticipated in the first stage of that program.

Notes

1. This section draws from A. G. Avarienova, "The Volga Basin and Its Regional Significance in Water Economy," *Vodnye Resursy*, No. 6 (1974). Also see A. B. Avokyan, G. P. Kalinin, S. L. Vendrov, and Yu. M. Magorzin, "Problems of Comprehensive Use of the Water Resources of the Volga Basin," *Vodnye Resursy*, No. 4 (1975).

2. Regarding river discharge, the second largest river in Europe is the Danube with its 200 km^3 of water annual discharge. Then follows the Pechora, with 130 km^3. In terms of river discharge the Volga is the fifth in the USSR, preceded by the Yenisei, Lena, Ob, and Amur (all four of them situated in the Asian USSR).

3. Except for the Neva River, its intra-annual runoff variations are regulated by lakes Ladoga and Onega, the largest in Europe.

4. The situation has changed with the Volga and Kama, having been regulated by a cascade of reservoirs; to date, April-June gives 25–38 percent of the annual volume and the other nine months gives 62–75 percent of it.

5. This section draws from S. L. Vendrov, *Problems of Modifying River Systems* (Moscow: Gidrometeoizdat, L.,

1970).

6. S. L. Vendrov, and K. N. Diakonov, "Storage Reservoirs and the Environment," *Nauka*, M. (1976).

3

The Azov Sea Water Economy and Ecological Problems: Investigation and Possible Solutions

A. M. Bronfman

Today when society is striving to extend, or at least maintain, its control over the environment, scientists increasingly turn their attention to creating natural-technological schemes that could have fixed parameters and functions. Most existing natural-technological systems evolved spontaneously due to the effects of human intervention in the environment. In the majority of cases the consequences of this intervention process were rather undesirable. To study these consequences, and to classify and predict them is extremely important in order to achieve enduring utilization of natural resources.

For some river systems this task has been carried out sufficiently well, but experience in tackling it for river-sea systems has so far been very poor. Intensive economic transformation of seawater resources results in many ecological effects which have been investigated only fragmentally. We do not have a full outline of those effects, though the need is great.

The discussion in this chapter is based upon the results of research carried out over many years in the Azov Sea. The choice of this basin was not accidental. In the first place, the environmental monitoring carried out in the Azov Sea area for

A. M. Bronfman is professor, Azov Scientific Investigation Institute for Fishery Economy, Rostov-on-Don, USSR. The outline given in this chapter was first presented to the symposium on "Man and Environment" (1976).

more than ten years has provided a rather complete set of empirical data. These data throw light upon various aspects of the interacting economic and natural components of the basin. In addition, the region lends itself to a modeling of its natural processes, thereby permitting thorough studies of the environmental consequences of human activity. Within the boundaries of the Azov basin it is now possible to outline a natural-technological system that is already partially monitored and controlled. The experience generated in forming this system is undoubtedly of great interest far beyond the outcomes for the Azov region itself.

The major natural and historical reasons commending the Azov Sea basin as a model are four in number.

1. The Azov Sea basin embraces interactions within one of the country's larger industrial-agrarian complexes, and constitutes a marine ecosystem of unique productivity. Under favorable sea conditions, the fish catches in the Azov Sea were sometimes as high as 9 t/km^2, and their potential is thought to reach 20 t/km^2. Such catches are unprecedented on the entire planet—either for sea or fresh water basins. The conflicts between social production and environmental protection are extremely pronounced.

2. The Azov basin is known for its rather limited river runoff resources, though the available waters are used intensively. The average annual drainage into the sea is only 41 km^3. The runoff coefficient for the catchment area is only 0.13. This coefficient is the lowest of all other large river systems in the European slope of the Atlantic. In 1975, the complete water consumption in the basin was 23 km^3, the irreversible consumption being 15 km^3, or about 35 percent of the mean.[1] The latter value is one of the highest in the USSR.

3. The anthropogenic transformations of Azov basin runoff have a long history, beginning in 1952 with the completion of the Tsimlyansk Reservoir on the Don River, having more than 10 km^3 in volume. Only the Rybinsk, in the upper Volga reaches (commissioned in 1947) is larger.

4. The Azov Sea ecological system is distinguished by its exceptionally slow development and its stability against external effects. The system's inertness in respect to its physiochemi-

FIG. 3–1

A GALOPATIC SCHEME OF THE BASIC TYPES OF AQUEOUS FAUNA IN THE AZOV–BLACK SEA BASIN
(according to Nordukhai–Boltovsky, 1953)

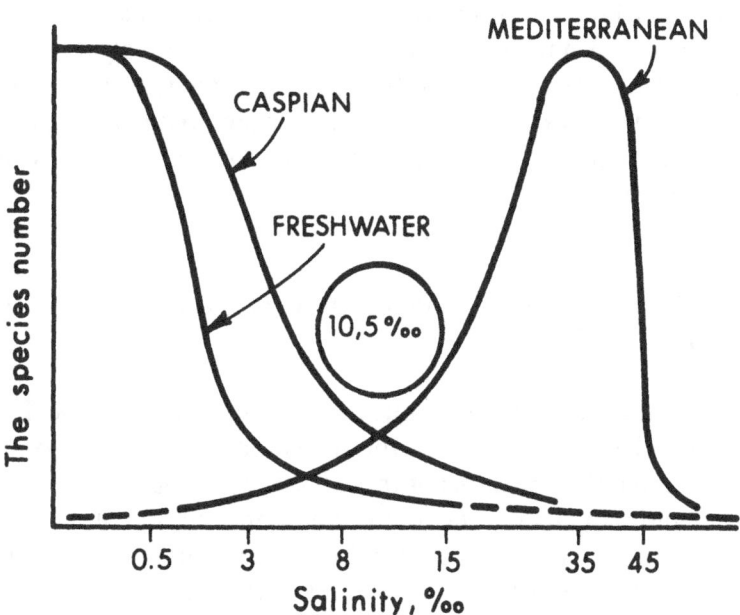

cal parameters is determined by the morphostructural features of the Azov Sea, and, primarily, by its rather small dimensions (surface=38,000 km², Hav=8.5m, volume=320 km³). The biological processes in the basin are characterized by rather short life cycles for most of its hydrobionts. The working reliability of any complicated system is known to be based on a general cybernetic principle of elemental abundance. Under this principle any defective element or interrupted connections can be duplicated many times. For natural communities this possibility is provided by a great diversity of flora and fauna. In the Azov Sea the well-known "saline waters paradox" is pronounced. The waters, despite their poverty in species, are usually highly productive. As seen from Figure 3-1, the Azov ecosystem, which evolved with an average sea salinity of about 10.5 percent, now has a very small number of species. This is why the system is so vulnerable to change.

The anthropogenically reduced river runoff and its intra-annual leveling by the upstream reservoirs have radically changed the conditions under which the Azov Sea ecosystem previously existed. Recently, it has been transformed into a system which can no longer utilize the material and energetic resources of photosynthesis as fully and effectively as it did in the past. Conditions for fish propagation within the Azov basin hydrographic network have been disturbed, and the environment for fish populations in the sea itself has been transformed.

Problems of Water Economy and Ecology in the Continental Part of the Basin

In 1975 the water was put to its highest level in the Tsimlyansk Reservoir and in the Krasnodar Reservoir on the Kuban River. The dams of these reservoirs hindered the routes for egg-laying fish in the middle and upper stream of the rivers. At the same time, the modulation of seasonal runoff sharply curtailed the frequency, area, and duration of floods in egg-laying areas of the floodplains situated in the tail races of the hydroelectric plants. With the Tsimlyansk Reservoir completed, the frequency of flooding in the lower Don egg-laying areas dropped from 84 to 18 percent of the years. The total flooded area dropped from 95,000 to 30,000 ha, and the duration of flooding from forty-nine to twelve days.[2]

In general terms, the ecological effects of regulated river runoff are well known[3] and, therefore, their detailed discussion in this paper is not necessary. Rather, it is important to speak about constructive solutions to those problems.

Two alternatives aimed at restoring the desired levels of reproduction of food-fish populations are under trial in the Azov Sea today. They are (1) further promotion of commercial fish breeding, and (2) specially regulated water discharges in order to provide optimal hydrological regimes in the tail waters of the reservoirs during the entire spawning cycle.

The first task has been almost completely achieved. In 1976, the fish-breeding plants of the basin released into the Azov Sea over 5.5 billion fries of the most valuable aboriginal species. This is about half of the total commercial fish husbandry of the USSR. The results are reflected in the restored stock of sturgeon, whose numbers over the last decade have enlarged fifteen fold. However, commercial fish breeding has so far only helped to slow down the fading of the populations of fish species which are less tolerant of the changed environment.

Every year spawning water discharges, following the recommended hydrographs, are directed into the lower Kuban reaches. In view of the precarious water-economic balance in the Don basin, these discharges can be utilized only about 30 percent of the years. They will be more frequent in the future when four low-head hydroelectric stations are completed in the lower Don. Then there will be a continuous system of dams and reservoirs and the bulk of the runoff, which has so far been used for maintaining the navigable depths in the river, will be managed for the sole purpose of supporting the fish economy. Two of the low-head stations are in operation, and the other two are to be commissioned in the near future. The eternal struggle between fish economists and hydroengineers, with the former defending the interests of fish farming and the latter those of the electric power stations, no longer exists here. It was possible to meet the aims of both the fish and power industries thanks to the specially designed structures for fish release and the unusual structures which are now laid down on the river bottom during periods of mass fish migrations for spawning.

Reconstruction of water movement in the lower Don, however, is a palliative. It provides a positive water-economic balance in the basin for only the next ten years. Beyond this horizon it is planned to satisfy the increasing demand for water chiefly by transferring to the basin increasingly large amounts of Volga runoff. By the end of the century, these transfers are expected to increase from 5 to 20 km^3 per year. Taking into account the constantly decreasing level of the Caspian Sea, the Volga, as the future donor of waters to the Azov basin, should be compensated, possibly by a supplement to its runoff from the northern rivers. Thus, the problems of the Azov Sea water economy go far beyond their territorial boundaries. The problems have become part of the general program for transferring the runoff of the rivers in the European USSR.

The suggested actions may eliminate the deficit in the water-economy balance of the Azov continental basin and guarantee desirable levels of natural fish propagation within that basin, but they cannot solve the entire problem of the sea. Indeed, no complete solutions are possible because of the adverse phenomena evolving in the sea itself.

New Processes in the Sea and Possible Ways for Their Optimization

In recent years the anthropogenic reduction of Don River runoff has been considerably aggravated by the decreased total moisture content of the basin. Reduced stream flow has increased the advection of the Black Sea waters, causing some 60 million tons of salts to accumulate annually in the Azov Sea. As a result, by 1976, the average salinity of the Azov Sea had increased to 13.8 percent (against 10.5 percent previously) and in the Don estuary to 10 percent (against 6.5 percent in 1912–1951). These changes in the Azov Sea salinity are illustrated in Figure 3-2.

The consequences of this salinization are manifested in many ways. First and most evident, they are seen in the sharply reduced lifespace for the brackish water and relict aborigines of the sea. Their range, confined to the 10–11 percent isohalines, covers only 10 percent of the sea area and constitutes 5 percent of its volume.[4] The ecological recesses of the remaining part of the sea, which became vacant, are now intensively filled by Mediterranean immigrants whose productivity is essentially lower.

These consequences represent only one episode in the long and branched chain of transformations in the sea ecosystem. The Azov Sea investigations make it possible to trace the nature of the deformations and establish a long series of changing patterns. So far, neither directly nor in any other way, has their appearance and evolution been related to the river runoff transformations.

FIG. 3–2
SPATIAL DISTRIBUTION OF SALINITY
IN THE AZOV SEA. A: 1931–1951, B: 1976

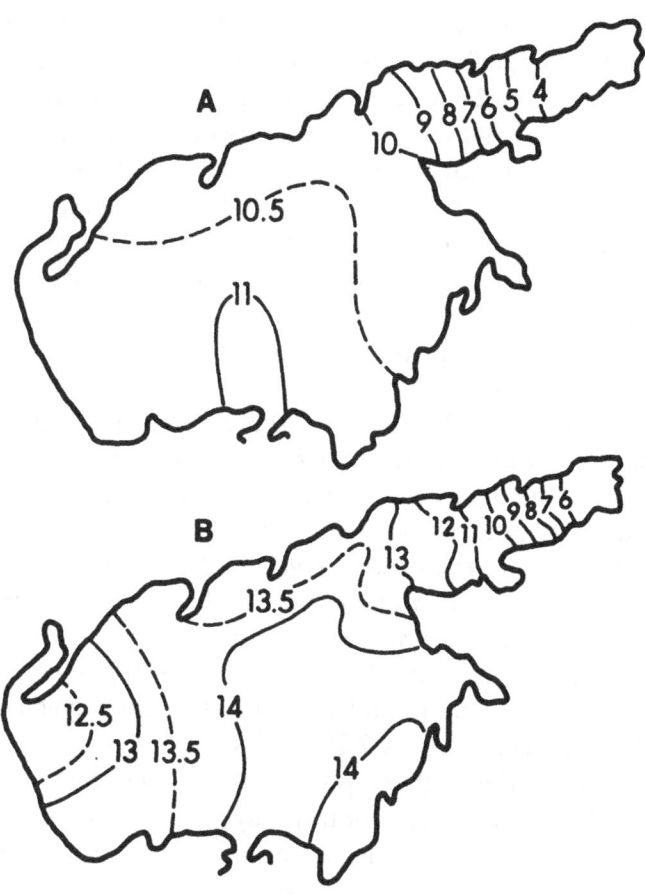

The active advection of the Black Sea waters greatly heightened the salt content and, consequently, the temperature
stratification of the seawater. In the six years between 1969 and
1975 the average annual values of its vertical stability have
increased from 1,140 to 3,500 conductivity units. The extreme

stability reached 115,000 conductivity units, while previously recorded figures never exceeded 36,000 units. In recent years, the intense density stratification of the Azov waters has never been violated even though the winds are strong (Table 3-1). Taking into account the shallow depths of the Azov Sea, this phenomenon can be regarded as one of the most recent. For comparison, the highest stability values in the Barents Sea never exceed 6,000 units, in the middle and southern Caspian they never go above 20,000 units.[5] In the vicinity of the Newfoundland Banks, even in the sections crossing the main line of the Labrador Current, they are 18,000 units.[6]

The weakened vertical water exchange has accelerated sedimentation of suspended organics and prompted their accumulation in the bottom sea deposits (estimated at from 1.63 to 2.55 percent). The increased organic mass has, in turn, enlarged the bacterial populations on the bottom by several orders of magnitude and, hence, the biochemical oxygen demand (BOD) in the surface ground layer (Table 3-1).

The somewhat passive character of the present dynamic aeration in the near-bottom waters under the increased BOD conditions is one reason why stable anaerobic situations or those similar to them can develop. For example, on the average, during the summer periods of 1961–1975, the vertical stability in zones with 80–100 percent oxygen saturation was 1,620 conductivity units. In those with 60–80 percent saturation it was 3,837 units and in zones with an oxygen deficit (under 60 percent) it was 7,290 units.

Because of the high trophic status of the Azov Sea, an oxygen deficit in the near-bottom layers could have existed in the past as well. If it did exist, it was only episodic and confined to a small area. During the recent ten to fifteen years these situations have been observed practically every summer, the deficit zone covering, on the average, 10 km^2 and in some years up to 20,000–22,000 km^2.[7]

Oxygen starvation of the Azov Sea has caused the periodic appearance of toxic products. Ground organics, anaerobically decayed (hydrogen sulfide, methane, phenols, carbonic acids, phosphine, etc.) have been found. This phenomenon can be called sea self-pollution. It should be studied thoroughly

FIG. 3–3

RATE CONSTANTS OF THE DECAY OF OIL PRODUCTS, SYNTHETIC SURFACTANTS AND PESTICIDES
VERSUS THE SEA WATERS SALINITY: 1. DIESEL FUEL, 2–3. SOLAR OIL, 4–5. NONIONIC
SYNTHETIC SURFACTANTS, 6. ANIONIC SYSTHETIC SURFACTANTS, 7. PROPANIDE

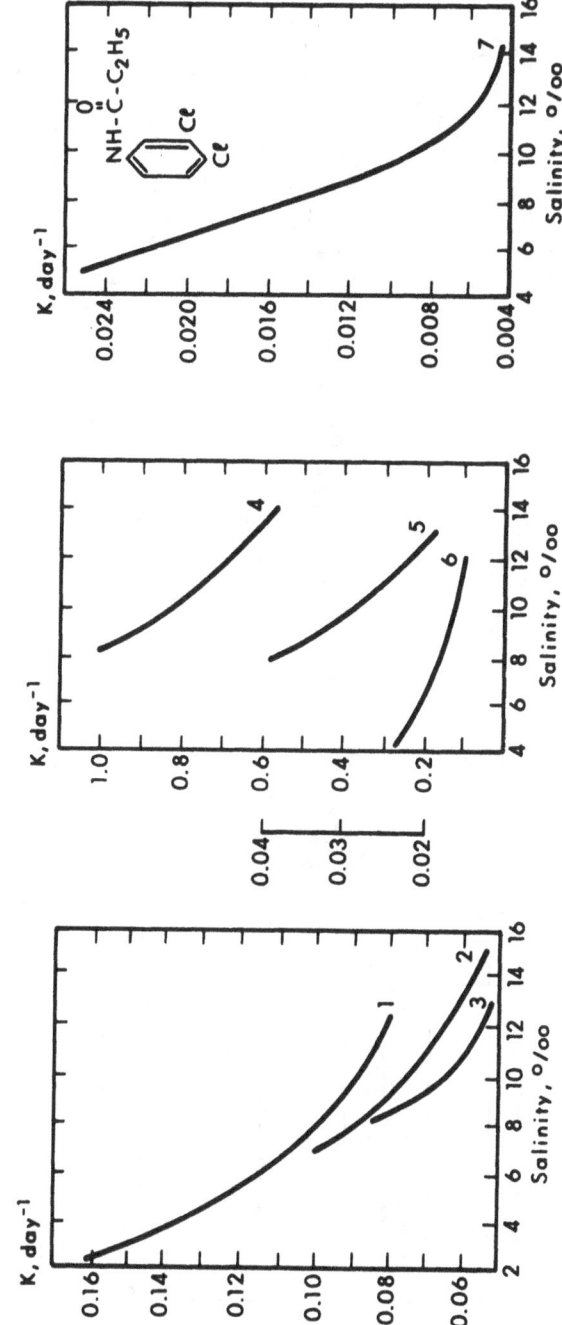

48

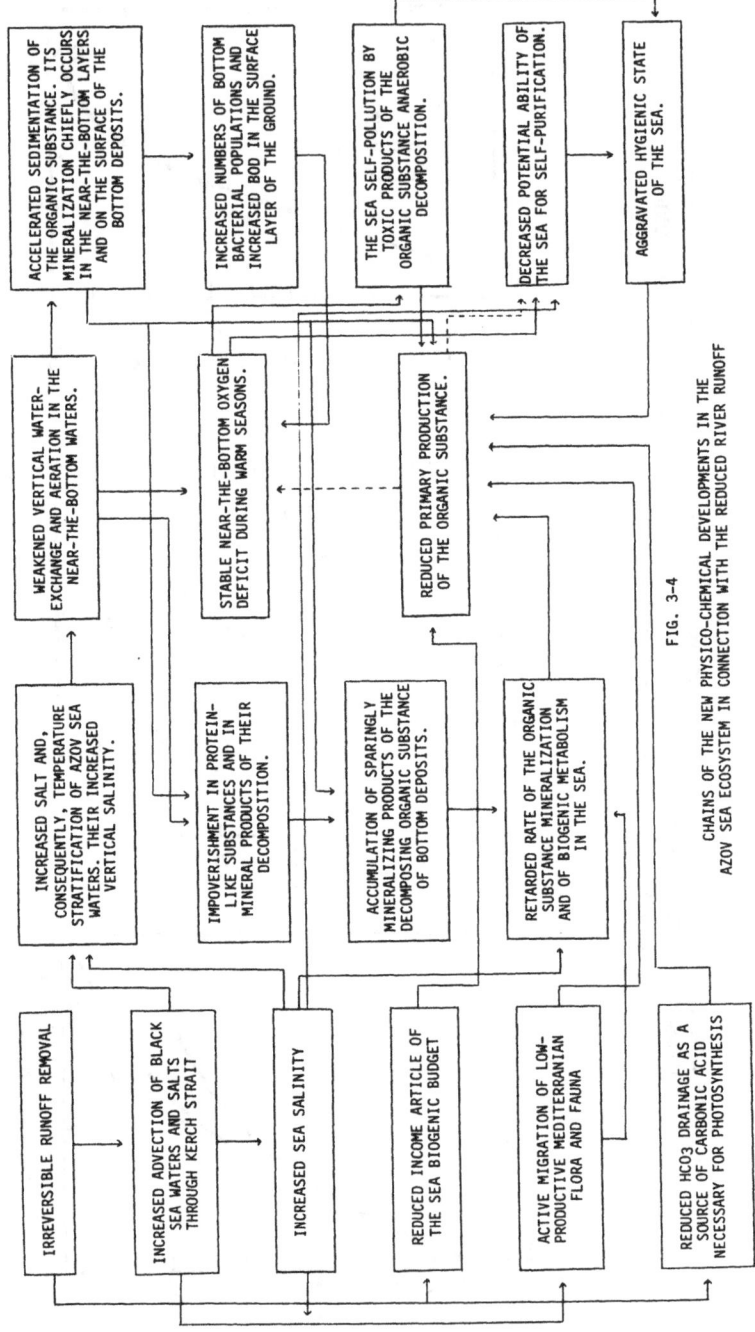

FIG. 3-4

CHAINS OF THE NEW PHYSICO-CHEMICAL DEVELOPMENTS IN THE
AZOV SEA ECOSYSTEM IN CONNECTION WITH THE REDUCED RIVER RUNOFF

inasmuch as the conditions contributing to its development exist in other seas as well.

Another equally serious consequence of the deteriorated oxygen regime of the Azov Sea is its decreased potential ability for self-purification from organic pollutants.[8] There may be another reason for this, directly related to the basin salinization. The experimental research thus far has shown that the salinity value is a reverse function of the decay rate constants for oil products, detergents, and pesticides (Figure 3-3).

The recent high salinity of the Azov Sea has lowered the rate of natural organic mineralization and, consequently, retarded the biochemical cycles of the main biogenic elements. As compared to the period of 1956–1960, when the average sea salinity was still rather low, the average rate of the nitrogen cycle by 1971–1975 had slowed down from 4.3 to 1.9 cycles per year. The phosphorus cycle has slowed from 12.1 to 5.7 cycles per year. The physiochemical reasons for this phenomenon are most probably connected with the increased ionic power of the solution.[9]

A monopolic development of diatoms is caused by the Azov Sea salinization. These diatoms conserve large organic masses during the decay of the siliceous algae shell. Phenomena of this kind, with final results very similar to those in the Azov Sea, have been observed by Rheinheimer.[10] He noted retarded decay processes in protein-containing compounds as well as ammonification, nitrification, and denitrification in some more salinized sea zones.

One reason for this reduced internal metabolism of biogenic elements that should not be neglected is the qualitatively transformed composition of the Azov Sea organics. The hypothetical explanation for this is related to the recent stagnation of the seawaters. Organic mineralization now develops mainly in the near-bottom layers and on the surface of the bottom deposits. Under these conditions, the pelagic sea area has lost many of the mineralizing products of the degraded planktonic organisms. In particular, it is no longer rich in proteins, carbohydrates, and lipids. The Azov Sea protein index,[11] calculated as a ratio between the protein-like nitrogen compounds and the total organic content, has dropped from

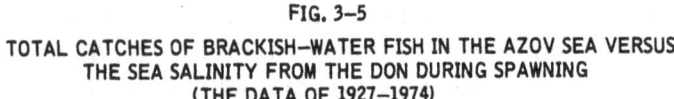

FIG. 3–5

TOTAL CATCHES OF BRACKISH–WATER FISH IN THE AZOV SEA VERSUS
THE SEA SALINITY FROM THE DON DURING SPAWNING
(THE DATA OF 1927–1974)

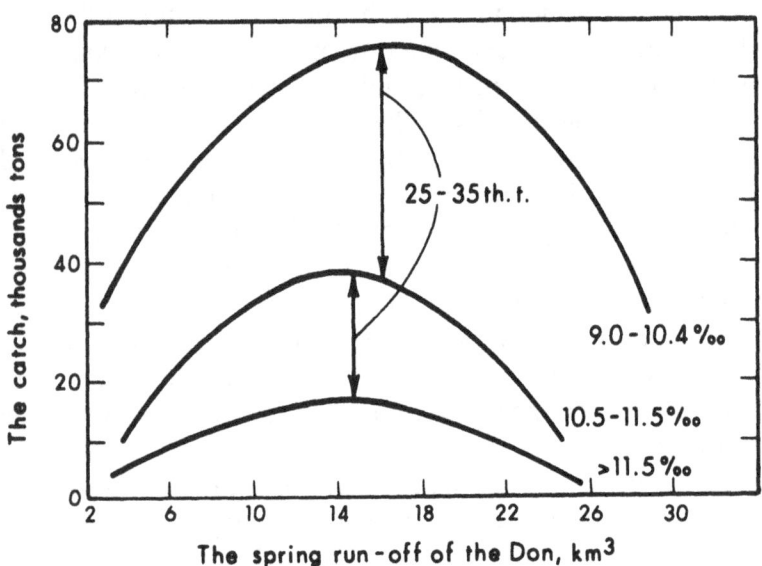

0.039 to 0.012. The organic substance deposited in the benthic zone is subjected to considerably deeper enzymic, microbial, and chemical transformations than that in the pelagic zone. As a result of these transformations, not only do mineral products develop, but some complex, slightly mineralizing organic compounds containing nitrogen and phosphorus (belonging to the humidic-guanidine group) also develop. Their accumulation in the seawaters greatly retarded the internal metabolic rate and decreased the trophic value of the organic substance. In the period 1971–1975 alone, when the average sea salinity rose from 11.8 to 13.2 percent, the ratio between the total bacterial numbers (billion cells) per mg of the organic substance decomposed daily lessened from 6.5 to 2.3.

These chemical transformations of the Azov Sea have

resulted in a greatly decreased content of the primary organic substance synthesized by its waters. During the last five years the amount of the organic substance annually produced by phytoplankton was 13–20 million tons, while the natural rate was 34 million tons. Among other reasons for the recent decrease in primary organic substance production the following certainly play a role: the changed specific composition of the sea phytoplankton; and the anthropogenic reduction of the biogenic river runoff and of hydrocarbonates (HCO^3). The hydrocarbonates constitute one of the important sources for the photosynthesis of carbonic acid.

The principal chains of cause-effect relationships which have evolved in the sea as a result of the anthropogenic reduction of the river runoff, are diagrammed in Figure 3-4. The scheme is not presented as final and absolute. It incorporates the first findings of the Azov Sea research. These findings, while tentative, permit two important conclusions.

1. Because of river runoff reduction, the sea ecosystem becomes damaged, not partially but as a whole. These damages can be found at different levels in the entire system.

2. Because of its structure, the Azov Sea ecosystem can be classified as a centralized one, where an element or a subsystem can act as a coordinator. This coordinating factor is the sea salinity. All important oceanographic sea parameters are consecutively connected, and correlated to this coordinating factor. At the same time, salinity is the main defective factor of the Azov ecosystem. The coincidence of these two properties helps explain the exceptionally fast transformation of the Azov Sea ecosystem. It also helps elaborate the most effective correcting measures.

The leading role of salinity is also evident in the resultant productivity of the Azov Sea. As seen from Figure 3-5, with its average mineralization increased by 1 percent, the annual catches of the most numerous brackish-water fish have diminished by 25,000–35,000 tons. In addition, reduced river runoff during spawning is estimated to decrease fish catches by 3,000 t/km^3.[12]

The Azov Sea salinity is the Gordian knot which must first be cut if significant improvement in the physiochemical and

Table 3-1

AVERAGE VALUES OF STABILITY, BIOCHEMICAL
OXYGEN DEMAND IN THE GROUND AND OXYGEN
CONTENT IN THE NEAR-THE-BOTTOM LAYERS IN
THE AZOV SEA*

	In the Zone of the Black Sea Waters Advection	Beyond the Advection Zone
(Winds Under 5 m/sec)		
Stability, con. units	8980	1546
BOD, g O_2/sq m /day	4.46	3.03
oxygen, ml/l	5.01	6.27
(Winds Above 5 m/sec)		
Stability, con. units	5183	883
BOD, g O_2/sq m /day	3.96	3.27
oxygen, ml/l	5.11	6.07

*According to the data of 1967-1975

biological state of the basin is to be achieved. There are, in principle, two practical ways to solve this problem. The first one would fully compensate the Azov Sea's irreversible water consumption by transferring river runoff. By the end of the twentieth century this consumption is likely to reach 30–35 km³.[13] Although this way is preferable because of its smaller ecological risks in the basin, its technological and economic aspects make it infeasible in the near future.

The salinity can be most effectively changed by a structure regulating the water exchange between the Azov and Black seas. This is the second alternative. It would be effective because the inflow and outflow of salts through the Kerch Strait make up about 97 percent of the income and expenditure items of the Azov Sea salt budget.

A statistical analysis of data for many years has shown that the basic oceanographic parameters of the Azov Sea become optimized in the salinity range of 10–11 percent. If the water exchange were properly regulated, this salinity level could be reached, on the average, within seven to ten years. It is only if there were extremely long periods of low river runoff that the transformation time might extend to fifteen years.

The second alternative has an important advantage. When realized, it might guarantee the optimal physiochemical conditions and, consequently, a high biological productivity for the sea, even if the river runoff were to be anthropogenically reduced by 18–20 km^3. With free water exchange, the maximum permissible irreversible removal, beyond which the sea ecosystem will inevitably degrade, never exceeds 5 km^3.

So far, the schedule for completion of this grand program has not been set. Because of the adverse phenomena rapidly developing in the sea ecosystem, we believe it is very important to solve this task as soon as practicable.

To regulate the water exchange between the two seas is an unprecedented practice which should be throroughly studied in all of its aspects. Serious attention should be devoted to the prediction of any unforeseen outcomes. At present, various details of this program are under intensive research. The aim is to obtain qualitative empirical data about the sea ecosystem, and to supplement existing measurement techniques with new ones, more adequate for quantitative analysis and prediction.

One step already yielding some promising results is a simulation model of the Azov Sea. This model can simultaneously display some 120 vector components of the sea's ecosystem. Structurally, it is composed of blocks and includes sixteen rather autonomous submodels.[14] The latter show the formation regularities and dynamics of the physiochemical and biotic elements of the basin (salinity, biogenic substances, oxygen, bacterio-, phyto-, and zooplankton, benthos, different fish populations, and so on). Presently, some individual blocks have been aggregated into an integral computerized system which has been tested for several retrospectives. The results obtained were sufficiently accurate to permit the model's further use as an instrument of prediction: it is believed to adequately reflect the intricate polyfunctional structure of the Azov Sea.

Some Aspects of the Clean Water Problem

A practical program for preventing qualitative depletion of the basin water resources is being implemented by special measures. The methods include closed, waterless industrial production cycles, decreased water consumption and water allotting rates and norms, and intensive construction of water treatment facilities. The latter involve some advanced facilities with new biological techniques for waste decontamination. The use of toxic or highly resistant pesticides (DDT and its homologs, heptachlorine, hexachlorine, polychloropinene) has been absolutely prohibited in the basin. These chemicals are now replaced by readily degradable preparations whose selective effects do not extract biologically active metabolites.

Following the requirements of the International Convention on Preventing Sea Pollution, ships in the Azov Sea have completely stopped their discharges of oil-containing wastes into the sea and the lower reaches of the basin rivers.

A wide system of controls and sanctions has been introduced in the Azov Basin. These measures are necessary for the implementation of the water protection program. The work already has had a positive impact. Such typical pollutants as heavy metallic salts, oil products, detergents, and pesticides are practically nonexistent in the Azov Sea. Where they do appear, their concentrations are within permissible limits (Table 3-2).

As seen from Table 3-2, the heavy metals concentration in the sea's pelagic zone is at the natural geochemical background level. This is much lower than the maximum permissible concentrations allowed in the USSR. The basin's health with regard to these types of pollutants is shown by the rather low content of heavy metals in the sea bottom sediments and by their lack of tendency to accumulate there.

Chloro- and phosphoro-organic pesticides are virtually absent in either the seawaters or in sediments. These chemicals have been found only episodically in some local zones strictly confined to the coast which now directly receive the drainage of irrigation waters. Purity from pesticide pollution is suggested by the fact that these preparations have not been absorbed

Table 3-2

HEAVY METALS CONTENT IN THE AZOV SEA

Metals and Their Permissible Water Concentration (PWC) mkg/1		Content Limits in the Pelagic Zone	Average Content in sediments, %	
			1970	1975
Pb	100	traces-4	0.0014	0.0015
Hg	5	no data	$<3 \times 10^{-6}$	$<3 \times 10^{-6}$
Zn	100	1-3	0.0029	0.0026
Cu	10	0.1-10	0.0033	0.0030
Ni	10	2-8	0.0043	0.0041
Co	10	2-3	0.0016	0.0015

by the hydrobionts' muscles or organs.

At present, the annual average content of detergents and oil products in the Azov waters (1975) is 0.11 and 0.05 mg/1 respectively. This is believed to be at a level permissible for fish-producing basins, and below the hygienic standards fixed in the USSR.

The results gained to date in correcting anthropogenic effects give grounds for optimism. Nevertheless, the planned development of the basin's economic potential dictates further measures protecting Azov water resources from qualitative depletion. Further improvement in measures for water protection should take into account two interrelated concepts.

First, the maximum permissible discharges of pollutants (both direct and carried by river runoff) should be at least one order of magnitude lower than the sea's total potential capacity for self-purification of its waters and of the sediments. If realized, this principle could be a guarantee of sea purity. Unfortunately, no quantitative assessment of the ratio is available today. Research in that direction hopefully will provide some solution to this cardinal question within a few years.

Water and waste treatment plants are nonproductive components in materials production. Their construction and use always involves increasingly heavy capital investment. Meanwhile, waste-water treatment can reduce significantly the toxic and harmful substances flowing into basins but cannot eliminate them completely. Today, it must be admitted that this is not yet within human power.

Second, because part of the pollutants will inevitably enter the basins, in choosing the optimal strategy for water protection, special attention should be paid to planning new technical projects and elaborating new measures which enhance the basin's capacity for self-purification. Such an approach is justified both economically and ecologically. According to calculations (so far only approximate), the total potential of the Azov Sea ecosystem, when including technogenic waste treatment, would produce some 500 million rubles. Klapper has made similar computations for alluvial channels in the Elbe River.[15] The expenditures there amounted to some 50 million marks per year. It is hardly logical to neglect such a mighty and effective natural mechanism. The only problem is to see that, in practice, it contributes to the basin's biological prosperity rather than causing adverse ecological effects.

The recent salinization and oxygen starvation of the Azov Sea has resulted in a degraded capacity for self-purification. The negative outcomes of this phenomenon are evident. They can seriously deteriorate the hygienic characteristics of the sea even if the present-day levels of discharges are maintained or slightly decreased. In this context, realization of the discussed complex of scientific measures aimed at optimizing the physio-chemical and biological relationships of the Azov Sea will at the same time facilitate the urgent task of maintaining high water quality.

Notes

1. A. M. Bronfman, "Runoff Transfer and Its Role in the Complex Solution of the Water Economical and Ecological Problems of the Azov Basin," in the collected papers, *Vliyaniye mezhbasseinovogo pereraspredeleniya stoka naprirodnye uslovia yevropeiskoy territorri i Sredinnogo regiona SSSR* (Moscow, 1975), pp. 71-81.

2. V. G. Dubinina, and Yv. M. Gargopa, "Fish Economy of the Azov Sea Basin Under Conditions of Intensive Water Resources Management," *Trudy Vsesoyvzngo nauch-issled, in-ta morskogo rybnogo Khoz-va i okenografii,* Vol. 103 (1974), pp. 10-13.

3. V. I. Vladimirov, P. G. Sukhoivan, and N. S. Bugai, "Fish Propogation Under Conditions of Regulated River Runoff," *Izdatelstvo A. N. UkSSR* (Kiev, 1963), p. 385; and S. L. Vendrov, "Problems of Transforming Some River Systems," *Gidrometeoizdat* (Leningrad, 1970), p. 236.

4. A. M. Bronfman, "The Azov Sea Salinity and Its Forthcoming Changes," *Investiya Serero-Kavkazkogo nauchnogo tsentra vysshey shkoly,* series *Estestvennye nauki,* No. 1 (1973), pp. 19-25.

5. *The Caspian Sea* (Moscow: University Publishers, 1969), p. 263.

6. A. A. Yelizarov, "On Vertical Stability of Water Masses in Fish-Producing Regions of Newfoundland Banks," in the collected papers, *Sovyetskiye rybokhozyaistvennye isslyedovaniya v severo-2ap. chasti Atlant. okeana* (Moscow, 1962), pp. 47-59.

7. A. M. Bronfman, G. D. Makarova, and M. G. Romova, "Recent Climate-Conditioned Changes in the Quantitative and Qualitative Composition of the Azov Sea Organic Substance," *Izvestiya AN SSR,* No. 6 (1973), pp. 39-48.

8. A. M. Bronfman, Kh. Ya. Zakiyev, and G. D. Makorova, "On the Question of the Azov Sea's Potential Ability of Self-Purification from Organic Pollutants," in the collected papers, *Okeanologicheskiye aspekty samoochishcheniya morya* (Kiev: Naukova duma, 1970), pp. 39-47.

9. A. M. Bronfman, and Yv. A. Dombrovsky, "Ecological Regularities and Statistical Models in the Formation of the Chemical Bases of the Azov Sea Productivity," *Izvestiya Severo-kavkazskogo nauchnogo tsentra vysshey shkoly,* series *Estestvennye nauki,* No. 1 (1975), pp. 71-77.

10. I. Rheinheimer, "Bakterion im Stickstoffkreilauf des Moeres," *Okosystemforschung* (Berlin: Hrsg. Heinz. Ellenbera, e.a., 1973).

11. Bronfman, Makarova, and Romova, "Recent Climate-Conditioned Changes of the Azov Sea," pp. 39-48.

12. A. M. Bronfman, G. V. Dubinina, and M. K. Spichak, "Quantitative Assessment of Some Ecological Outcomes of the Anthropogenic Activity in the Azov Sea," in the collected papers, *Rybokhozyastvennye issledovaniya v basseine Azovskogo morya,* Rostov-on-Don (1972), pp. 27-38.

13. Bronfman (1975), "Runoff Transfer and Its Role," pp. 71-77.

14. A. B. Gorstko, L. V. Abetsedarskya et al., "The Azov Sea Simulation System," in the collected papers, *Metody sistemnogo analiza v problemakh ratsionalnogo ispolzovaniya vodnykh resvrsov* (Moscow, 1976), pp. 81-162.

15. H. Klapper, "Zur Belastbarkeit und Selbstreinigungsleistung der Gewässer," *Wasserwirtschaft-Wassertechnik,* Vol. 25, No. 8 (1974).

4

The Problem of Transferring Runoff from Northern and Siberian Rivers to the Arid Regions of the European USSR, Soviet Central Asia, and Kazakhstan

I. P. Gerasimov and A. M. Gindin

Water resources in the USSR are being transformed profoundly under the impact of economic activity. Basic social, technical, and economic changes are realized in the interest of improvement of the human environment and of a more effective use of natural resources. The systems of large dams and reservoirs built on the Dnieper, Volga, Kama, Angara, and other rivers and those under construction on the Ob and Yenisei are examples of how the natural runoff may be modified by seasonal regulation, thereby providing great possibilities for the development of electric power, transport, and water supply.

The role of complex river development in the development of the production factors of Soviet society and in the completion of economic, social, and cultural tasks grows every year. Being a factor of production, water resources precondition the

I. P. Gerasimov is director, Institute of Geography, Soviet Academy of Sciences, Moscow. A. M. Gindin is assistant chief, Department of Nature Use and Conservation, State Committee of Sciences and Technology, Moscow.

development of certain regions of the country, the location of industries and population, and allow for radical improvement in the conditions for the people's working, living, and recreation.

A General Review of the Water Economy of the USSR

Renewable water resources in the form of river runoff from the territory of the USSR make up 4,358 km³ a year. The USSR has the largest total volume of freshwater resources in the world. However, the distribution of water resources within the USSR is extremely irregular and does not correspond to the distribution of population, industry, and agriculture. In the European USSR, where 75 percent of the country's total population is located, 27 percent of the nation's water resources are available. The greater part, 64 percent of the country's river water, flows to the basins of the Arctic Ocean. The four great rivers—Yenisei, Lena, Ob, and Amur—carry 44 percent of all fresh waters. Fifteen percent of the country's water resources are confined to the southern and western regions situated in the basins of the Atlantic Ocean and in closed drainage areas of the European USSR and Soviet Central Asia.

The development of the factors of production in the country, the growth of population and its culture lead to an increased consumption of water, which is illustrated by Table 4-1. By 1975, as compared with 1913, per-capita water consumption had increased five times. According to forecasts of economic development, by 1990 water consumption will continue to increase in all sectors of the economy.

The river runoff within the USSR, including the inflow from the territories of adjoining countries, amounts, on the average, to 4,714 km³ a year. It would seem that with such resources the needs of the country's economy for water should be fully satisfied. However, as noted above, water resources are unevenly distributed. The difficulties are aggravated by an unfavorable annual distribution of the runoff on the major rivers of the Soviet Union.

TABLE 4-1

WATER CONSUMPTION IN THE USSR: 1913, 1940, AND 1975

	1913	1940	1975
Population, million	159	194	253
Water intake, cu. km/year	45	80	342
Water consumption per capita, cu. m/year	282	415	1360

The north-south contrast is of basic importance. The southern zone of the country, where 80 percent of the population and 80 percent of the industrial and agricultural production are concentrated, has only 14 percent of all river water. Of those resources, 80 percent are from the following eleven rivers: Volga, Dnieper, Dnestr, Don, Kuban, Ural, Terek, Sulak, Kura, Amu-Darya, and Syr-Darya (see Figure 4-1). The economic water budget in the basins of these rivers is given in Table 4-2.

Continued growth in water consumption will lead to difficulties in supplying the economy with water in the southern USSR. Without special large projects for redistribution of the runoff, it may, by 1990, considerably lower the gravitational level of the Caspian Sea and, as indicated by A. M. Bronfman, seriously increase the salinity of the waters of the Azov Sea. According to fishing industry authorities, such changes in the level of the seas and their salinity may greatly decrease fish productivity.

A number of large projects have been undertaken to provide the Soviet economy with water. These included the construction of 3,740 km of canals providing for an interbasin transfer of 40 km³ of river runoff, and 205 large water reservoirs, with a capacity of 400 km³, for regulating river runoff. Ninety of the

FIG. 4–1
SELECTED USSR RIVERS

reservoirs were built near hydropower stations, their effective capacity being 361 km³. This volume increases the base runoff by 20 to 25 percent.

However, local water resources cannot meet all the requirements of the future economy of the southern USSR, especially those of irrigated agriculture. That is why one of the major problems confronting the country today is that of transferring the northern European and Siberian river waters to the arid and desert plains of Soviet Central Asia and Kazakhstan. These regions receive much more solar light and heat than the others and their frost-free period is the longest in the country, but their precipitation is minimal. These climatic conditions create acute problems for economic development in those areas. Water resources are placed at the head of all natural resources—as water is needed for domestic use, for industry,

TABLE 4-2

ECONOMIC WATER BUDGET OF
RIVERS IN SOUTHERN USSR

Basins	Water resources	Irreversible water consumption 1975
All rivers of the USSR, cu. km	4714	186
Eleven rivers of the southern slope, cu. km	520	115
Per cent of the total volume	11	61.5
Caspian Sea, cu. km	300	41
Azov Sea, cu. km	40	12.7
Aral Sea, cu. km	110	84

and for the development of irrigation and husbandry.

Against the background of this general outline, it is helpful to give detailed consideration to the problems of supplying water to one of the more water-deficient regions of our country—the basin of the Aral Sea. The Aral Sea basin includes the whole of Soviet Central Asia and southern Kazakhstan with a total area of 2.4 million km². It is one of the more important economic regions of the USSR and the largest and most ancient region of irrigated agriculture. About 50 percent of the country's irrigated lands (with a population of about 30 million people) are concentrated there. It has considerable land resources suitable for irrigation: a total area reaching 50 million ha; agroclimatic conditions favorable for cultivating heat-loving crops; labor resources with long experience in irrigation and a rapid growth rate; and well developed industry and transportation.

The development of irrigation and other water-consuming branches of the economy has turned some river basins of this territory into complex hydrological systems for managing water. Hydrological construction greatly advanced in the

Soviet period. Brought into operation were many storage reservoirs (Karakum, Chardarya, Chimkurgan, South Surkhan,
and others); hydropower stations; water aqueducts (Amu-
bukhar, Karshin, and others); the largest artificial river in the
world (900 km); as well as such sophisticated and up-to-date
irrigation systems as Golodnaya Steppe, Surkhan-Shirabad,
and many others. The Toktogul, Charvak, Nurek, and Tyuya-
muyun reservoirs and the large engineering system in the
Karshina Steppe are now being built. The total length of
constructed irrigation canals has reached over 250,000 km and
that of the intercepting drainage system about 90,000 km.
These measures made it possible to enlarge the total irrigated
area in the republics of Soviet Central Asia and south Kazakhstan to 6.3 million ha by 1975. During the four years from 1971
to 1975 more than 800,000 ha of new lands were irrigated and
used in agriculture. As a result of the vast irrigation construction, it was possible, even in the extremely dry year of 1974, to
obtain a maximum yield of cotton.

In the future, the republics of Soviet Central Asia will have
to solve a number of important and complicated economic
problems involving promotion of irrigated cultivation for
increased production of cotton, rice, vegetables, fruits, and
grapes, as well as for expanding husbandry. These will turn on
availability of water. Notwithstanding its large economic and
natural potential, the Aral Sea basin is deficient in water
resources. It has been estimated that at least 400 km^3 of water
are needed annually to irrigate the better land of this region. In
contrast, the basin's river runoff averages approximately 127.5
km^3 a year. The flow is 98.9 km^3 in dry years with 90 percent
rainfall probability.

The disparity between the annual runoff and water consumption leads to sharp deficits in dry seasons (repeating every
five or six years), and makes it necessary to regulate the runoff.
of the Central Asian rivers both seasonally and annually. By
careful regulation it may be possible to maintain a reliable
runoff of up to 100 km^3 in years of 90 percent rainfall
probability. In practice, some 90–92 km^3 may be used out of
this volume, a factor which is generally taken into account
when planning the economic development of the Aral Sea

basin. Of the rest of the river runoff, some evaporates from the reservoirs and rivers, some goes to foreign countries (Afghanistan and Iran), some consists of extreme maximum discharges uncaught by the reservoirs, and some is required to transport sanitary discharges.

With such water deficiency, a further development of the Aral basin economy is only possible if the local water resources are used rationally, and if a part of the runoff of the Siberian rivers is brought to this basin. To solve the problem of supplying water to this territory the following three major steps should be taken: (1) the runoff of the rivers of Soviet Central Asia should be regulated and redistributed in order to use effectively all the water resources of the basin; (2) existing irrigation systems should be remodeled in such fashion as to increase their efficiency and provide a more rational and beneficial use of the available irrigation water; and (3) a part of the runoff of the northern and Siberian rivers should be transferred to the Aral Sea basin.

Redistribution of River Runoff Within the USSR

Having reviewed tentative projects for redistribution of water resources, four different approaches may be outlined. Each of them solves one and the same problem—supplying the nation's economy with water for a period of at least 30–40 years.

The *first variant* assumes that the European and Asian parts of the USSR would be provided wtih water separately. The water deficit of certain regions in the European USSR would be covered by the runoff of the northern rivers of the Karsk and White sea basins and the lakes of the northwest of the RSFSR, which would be transferred to the Volga River. In addition, some water from the northwestern rivers would go to the Dnieper basin and some water of the Danube. The water needs of the Asian part of the country, mainly of Central Asia and Kazakhstan, would be satisfied by transferring a part of the Ob and Yenisei rivers' runoff from their upper and middle reaches.

The *second variant* proceeds from a joint solution of the

European and Asian water problems. The idea is to take water from the lower Ob and to transfer it across the Urals to the basin of the Pechora and further on to the Volga. The waters of some other northern rivers and lakes of the European USSR would be transferred to the Volga. The major water consumers in Soviet Central Asia and south Kazakhstan, middle and lower Volga river region, north Caucasus, Kalmykiya, and Rostov Oblast would be supplied from the Volga. Some other local variants of river runoff transfers are also suggested, among them the transfer of Danube water to the Dnieper, or of part of the northwest rivers to the Dnieper and elsewhere.

The *third variant* presumes that the runoff of the Volga river may provide water for all the major regions of the Asian and European USSR. As the volume of the Volga water coming to the Caspian Sea decreases, some of the Black Sea waters would be transferred to the Caspian Sea. A dam, to be built in the northern part of the Caspian Sea, is planned in order to control the water-salt regime and to connect the northern and central parts of the Caspian with a sluice. As in the first two variants, some local projects relating to other basins would also be included.

The *fourth variant* may be considered as a combination of all possible technical solutions realized on an optimum level and scale within a "single water economy system" (SWES) for the country. The last variant may be looked upon as an advanced project for a water system in the USSR, comprising a comprehensive solution of its water problems. The first three variants may be regarded as separate stages in the development of the SWES, the formation of which may take more than thirty or forty years.

It is believed that river runoff regulation should be accomplished not only by traditional methods of constructing intercontinental reservoirs but also by regulating water in the estuarine areas of rivers. To this end, sea bays, gulfs, and sections of sea coasts would be used.

The creation of regulating water bodies in the river estuaries within the basins of the Karsk, Barents, White, and Baltic seas, and the use of natural sea bays (the Ob and Yenisei Gulf, some bays of the White and Baltic seas), their waters transferred to the south, would make it possible to preserve the natural hydrological regime of the river systems and to do without additional reservoirs in plains terrain. By regulating the exchange between the resultant water bodies and those of the seas, it should be possible to provide suitable conditions for fisheries and sea transport. With a similar approach to the regulation of water resources in the south, water bodies with a regulated salt regime could be created. Such water bodies, receiving rich biogenic runoffs could provide conditions for intensifying the fishing economy in these warm territories of the country (the Azov Sea, the north Caspian, and the Black Sea bays).

Although each of the above variants plans approximately equal volumes of water consumption, they would each produce different impacts on the use of natural resources, the development of production, and the dynamics of natural processes. Some features of those impacts may be resolved during the planning and designing stages. Others—and these are a majority—require special research and forecasting in advance. For example, thinking in terms of regional water budgets, the projects of the first variant (with separate European and Asian transfers) will probably decrease streamflow in the lower reaches of the northwestern rivers of the European USSR and also in the middle reaches of the Ob. The second variant (the joint European-Asian transfer) makes it possible to avoid those effects.

The second variant has some other advantages in respect to the development stages and shorter freezing of capital investments during construction. For instance, with the construction of the Volga–Aral Sea canal underway and with the basin of Syr-Darya and Amu-Darya already provided with water, a beginning could be made in the transfer of water from the north to the Volga basin. The waters of the Pechora, North Dvina, the northwestern rivers, and the lower Ob could be used in different volumes and in different sequence. At this stage a change in the accomplishment of the third variant system is also possible, i.e., the transfer of the Black Sea waters to the Caspian Sea. Great volumes of water transferred to the Volga may improve the quality of the water in the Volga reservoirs as well as the entire river runoff of the basin.

The difficulties of transferring the lower Ob waters across the Ural Mountains are apparent. On their way they would have to traverse areas of permafrost. Also technically complicated would be the transfer of waters from the Black Sea to the Caspian as well as the regulation of the salinity regime of the north Caspian. A further complication is the problem of the salinity regime in the Aral Sea and the fact that the canal would be built across a highly developed region.

The fourth variant, based on an optimum combination and sequence of separate technical projects, would avoid some of the drawbacks of the other variants. An important benefit of this variant is the possibility of creating an SWES for the USSR which would establish the management of USSR water resources on a new basis. By setting up regulated water bodies in the estuary areas of the rivers and by transferring water over large territories, it would consider the disparities between the availability of water and its consumption rates.

A huge meridional trough connecting western Siberia with Central Asia may be regarded, at first sight, as favorable to the transfer of the Siberian rivers' runoff. At present this trough is occupied by the Tobol River (a tributary of the Irtysh) in the north and by the Turgai River in the south. In the geological past the Turgai probably flowed into the Aral Sea. It has been suggested that during the glacial period the trough was a drainage channel for the lake and river waters swollen by the melting northern glacier. Despite the fact that these speculations have proven to be wrong, the very existence of the Turgai corridor encourages people to look to possible ways of using it for artificial transfer of river waters.

The future destiny of the Aral Sea is an important part of the problem under consideration. The level of the basin depends upon the Amu-Darya and Syr-Darya runoff. As a result of complex river development and an ever-increasing use of surface water for irrigation, the water level in the Aral Sea is gradually lowering. In the last ten years it dropped by 3 m. Calculations have shown that if the whole runoff of the Amu-Darya and Syr-Darya rivers is used, the Aral Sea will turn into a dead residual basin and its low level will be maintained only by drainage waters. If, under the same conditions, the Aral Sea receives the Siberian water, there will be a possibility of maintaining its level and regime.

The grand scale of the problem—its close connection with the economy of the country, the expected impact on the environment due to unprecedented interference with nature, long-term and irreversible effects, and the initiation of certain natural processes—necessitates extensive research and special design work.

Taking all this into account, the 24th Congress of the Communist Party of the Soviet Union decided to plan research for 1976–1980 along those lines. By 1976, in accordance with that decision, the work program was being elaborated to enlist the cooperation of the major scientific and design institutions of the USSR so that their joint efforts might help solve the problems in the best possible way.

The tasks confronting scientists and engineers are extremely important. They must choose the most suitable way for the transfer as well as the optimal technical structures (canals, dams, reservoirs, pumping stations, etc.). They must determine the places, volumes, and sequence of water withdrawal, filtration, and evaporation losses. Their studies must help specify the most rational use of the Siberian waters for irrigation and for watering pastures. They must also make scientific forecasts of the physicogeographical consequences of future transfers. Such forecasts should include:

1. determination of the possible effects of the water withdrawal from the Siberian rivers upon the regime and bioproductivity of the coastal waters of the Arctic seas;
2. changes in the total moisture cycle on dry lands and in the atmosphere of western Siberia, Kazakhstan, and Soviet Central Asia caused by the redistribution of the river runoff;
3. possible changes in the water budget and regime of designated sections of the affected rivers, lakes, and groundwaters;
4. changes in the natural land ecosystems (relief, local climate, soils, permafrost, vegetation cover, fauna, and

population) in areas of the future transfer;

5. opportunities to preserve and improve aquatic ecosystems in water basins related to the future transfer (in particular, the resources of fish);

6. determination of the impact of water redistribution on the location and development of production factors and the living conditions of the population; and

7. changes in the natural and economic characteristics of the Aral Sea.

References

Gerasimov, I. P. "Problems of Natural Environment Transformation in Soviet Geography." *Progress in Geography: International Reviews of Current Research,* Vol. 9. London: Edward Arnold Ltd.

Gerasimov, I. P., et al. "Large-Scale Research and Engineering Problems for the Transformation of Nature in the Soviet Union and the Role of Geographers in Their Implementation." *Soviet Geography: Review and Translation,* Vol. 17, No. 4.

Vendrov, S. L. "Problems in the Spatial Redistribution of Streamflow." *Soviet Geography: Review and Translation,* Vol. 17.

5

The Columbia River

Keith W. Muckleston

This chapter has three goals: first, to describe impacts that water resource development has had on a broad spectrum of phenomena in the Columbia basin in the United States; second, to outline gaps in knowledge that need to be addressed; and, third, to identify research efforts that have contributed to a better understanding of the interrelationships between water resource development and the phenomena considered.

The following two sections provide a background by briefly describing the salient features of the drainage area and the major elements in the water resource development program.

Characteristics of the Drainage Area

The Columbia drainage area is approximately 812,000 km² (313,500 mi²). It comprises the Columbia–North Pacific Water Resources Region which covers 709,660 km² (274,000 mi²) and the Columbia and Okanagan drainages in Canada which add about 103,600 km² (40,000 mi²). The Columbia drains 670,530 km² (258,892 mi²), which comprises 83 percent of the study area. The average total runoff is 8,773 m³/s (310,000 f³/s), the

Keith W. Muckleston is associate professor of geography, Oregon State University, Corvallis, Oregon.

Columbia accounting for 6,764 m³/s (239,000 f³/s). The variation of momentary flows on the Columbia is about thirty-five to one. The average annual precipitation varies markedly, from approximately 5100 cm (200 in) to less than 13 cm (5 in); runoff varies from over 400 cm (160 in) to less than 1.3 cm (0.5 in). The Cascade Range divides the study area into a marine-moderated, humid subregion to the west and a much larger, more continental, semiarid subregion to the east.

The distribution of the region's 6.8 million people is uneven, characterized by relatively dense agglomerations in the Puget Sound–Willamette lowlands west of the Cascade Range and a generally sparsely populated interior. The western sector has been responsible for most of the regional increase of population, while decreasing population has been common in parts of the eastern subregion. The Canadian portion of the study area has approximately 282,000 inhabitants, many of which are concentrated in the Kootenay and Okanagan valleys, and along the Columbia near the international border.

The region has a relatively heavy dependence on the extraction and primary processing of natural resources. The associated economic activities are distributed at or near the principal loci of logging, agriculture, mining, and large hydroelectric plants. The majority of the labor force is employed by the tertiary sector which is generally located in the metropolitan areas.

Major Elements in the Development Program

A variety of tools are used in the region to produce a full range of water-derived services (Figure 5-1). Water development in the region is characterized by a heavy reliance on flow uses, particularly to generate hydroelectric energy. In accordance with evolving socioeconomic conditions, water uses associated with leisure time activities are becoming increasingly significant. A relatively small proportion of water is used off-channel. East of the Cascades, irrigation accounts for approximately 95 percent of the total withdrawals. In the more densely populated and industrialized western part of the region, public

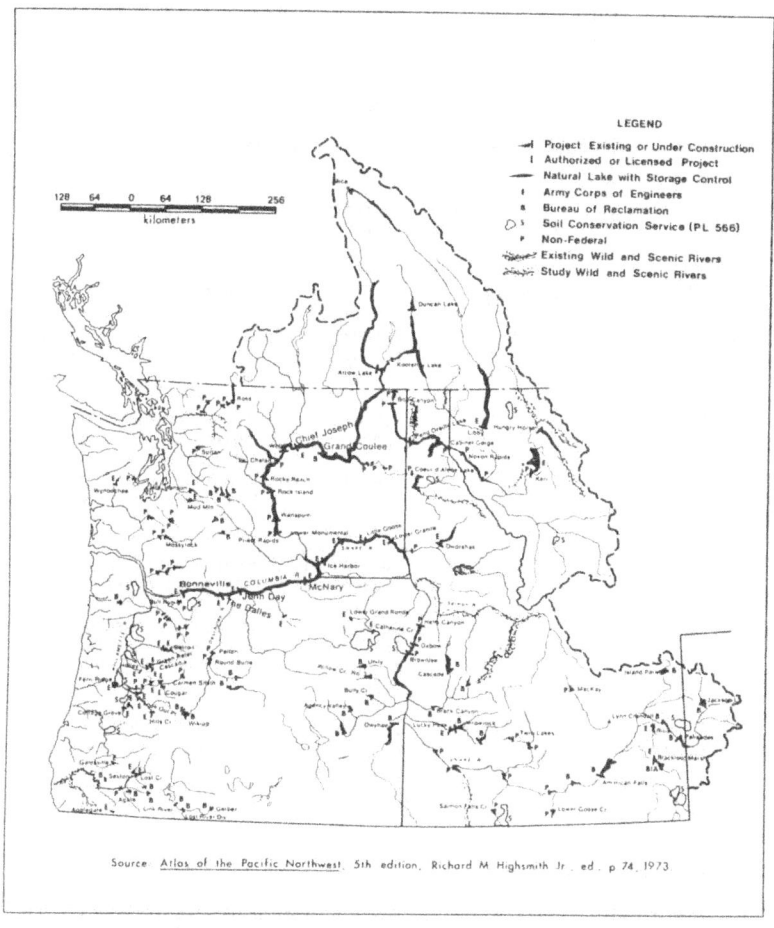

Source: <u>Atlas of the Pacific Northwest</u>, 5th edition, Richard M Highsmith Jr, ed, p 74, 1973

FIG. 5-1 WATER MANAGEMENT TECHNIQUES

water supply and self-supplied industry are responsible for approximately 70 percent of total abstractions. Pulp and paper mills account for most of these withdrawals.

Four of six strategies of water development are widely employed in the region.[1] A strategy is defined as a characteristic combination of goals, means, and decision criteria. Categorization depends on: (1) whether the goal is single or

TABLE 5-1

EMPLOYMENT OF WATER MANAGEMENT STRATEGIES IN THE REGION

Strategy I   Goal: single purpose   Means: construction
                  Decision Criteria: private

Water Derived Services	Use in the Region
1. Domestic Water Use	Dominant until 20th Century then supplanted by II-1. Still significant in rural areas.
2. Irrigation	Dominant until first decade of 20th Century. Continual employment to the present. Renewed vigorous expansion over the last decade, utilizing technological innovations and improved techniques of management.
3. Navigation	Dominant in pioneer phase of settlement: cleared waterways, built docks and supplied river craft. Became heavily dependent on Strategy II since end of the 19th Century and on III since 1930's for channel maintenance and lock operation, respectively.
4. Generation of Hydroelectric Energy	Dominant during first decades of the century then surpassed by Strategies II and III in 1930's. Have remained significant to the present--most of new generating capacity since 1970 is thermal.
5. Flood Hazard Adjustment	Limited diking activity--limited success.
6. Industrial	Dominant for industries with large water requirements. Dominant for industries with small requirements until 20th Century.
7. Waste Carriage	Dominant for industries with large volumes of waste (with increasing degrees of treatment by these industries). Wastes from industries with moderate to small volumes of waste treated increasingly under Strategy II.

Strategy II   Goal: single purpose   Means: construction
Decision Criteria: public

Water Derived Services	Use in the Region
1. Municipal (includes Domestic) and Industrial	Dominant supplier of households and industries with small water requirements since early 20th Century.
2. Irrigation	Dominant in developing new irrigated lands 1910 to 1940's. Supplementary supply to I-2 to the present.
3. Navigation	Dominant since late 19th Century for channel maintenance.
4. Generation of Hydroelectric Energy (Public Utility Districts)	Significant since 1930's. Much of the new generating capacity since 1970 is thermal.
5. Flood Hazard Adjustment	Some dike and levee construction. Levee system most significant along Lower Columbia.
6. Industrial	Industries with moderate water requirements supplied by Strategy II-1.
7. Waste Carriage	Dominant over last five decades for municipalities and for industries with moderate volumes of waste.
8. Fish Propagation	Carried out by State and Federal agencies in an attempt to mitigate losses of anadromous fish due to dam construction and other human activities.

Strategy III   Goal: multiple   Means: construction
Decision Criteria: public

Water Derived Services	Use in the Region
1. Municipal and Industrial	Relatively small volume of storage provided since passage of Water Supply Act of 1958.
2. Irrigation	Dominant in developing new irrigated lands late 1940's through 1960's. Provision of supplemental water to established irrigation systems developed under I and II.

TABLE 5-1 (continued)

3. Navigation — Significant factor in extension and improvement of inland water transport from 1930's to the present.

4. Generation of Hydroelectric Energy — Dominant producer since early 1940's.

5. Flood Hazard Adjustment — Dominant adjustment by flood flow modification since 1930's.

6. Industrial — Some storage for industrial use since passage of the Water Supply Act of 1958.

7. Waste Carriage — Dilution of pollutants by low flow augmentation since 1940's, although relatively little storage is allocated specifically for flow augmentation to facilitate waste carriage and assimilation.

8. Fish Propagation — Carried out by State and Federal agencies as part of multipurpose structures. Classed as either mitigation or enhancement, depending on the circumstances.

9. Recreation — Carried out singly and in combination by State and Federal agencies at multipurpose Federal projects. It has become particularly significant over the last three decades.

Strategy IV   Goal: single purpose   Means: multiple
Decision Criteria: public and private

Water Derived Services	Use in the Region
1. Flood Hazard Adjustment through flood insurance and land use regulations.	Limited implementation from passage of the first Flood Insurance Act of 1968 until 1974. More rapid implementation since 1974 when sanctions were added by the Flood Disaster Protection Act of 1973 (PL 93-234).

Strategy V   Goal: single purpose
Means: multiple, including research as a conscious management tool
Decision Criteria: public

<u>Water Derived Services</u>	<u>Use in the Region</u>
1. Weather Modification	Limited application in Northern Rocky Mountains in Montana by Bureau of Reclamation but no longer operational. Limited application by the State of Washington in the North Cascades. Some cloud-seeding in central Oregon.

Strategy VI   Goal: multiple
Means: Multiple, including research   Decision Criteria: public and private

<u>Water Derived Services</u>	<u>Use in the Region</u>
1. Multiple Services in Metropolitan Areas	Mechanisms not yet present in the region.

multiple; (2) whether the means used to achieve the goals are construction, regulation, research, or some combination; and (3) whether the decision criteria are established by the public or private sector. Table 5-1 outlines use of the strategies in the U.S. part of the region.

Domestic and Industrial Water Supply

Domestic water is supplied by 850 municipal systems to approximately 75 percent of the population. Thirty-four large systems supply about 53 percent of the regional population. While municipal systems supply industries with modest water requirements, industries requiring large volumes of water are self-supplied.

Irrigation

Because the region is characterized by dry summers and

much of it is semi-arid, irrigation has been used extensively to increase production. Approximately 3.1 million ha (7.7 million ac) are irrigated. The most extensive areas of development are in the Snake River plain of southern Idaho and in the Yakima Valley and Columbia basin projects, both of which are in the central part of Washington. Although there is more irrigated land in Oregon than in Washington, it is scattered among many small developments. During the past several years, agribusiness has brought much of the newly irrigated land into production.

Electric Generating Capacity

In 1975 the regional generating capacity (excluding that in Canada) was 25,527 mw of which about 82 percent was hydropower. That year hydroelectric plants in the U.S. part of the region utilized over 1,726 km^3 (1.4 x 10^9 ac ft) of water which exceeds the total runoff many fold. In 1975 an additional 12,177 mw of generating capacity was under construction, 61 percent of which was hydroelectric. Much of the additional hydroelectric energy is being developed by installation of more generators at existing dams. At Grand Coulee, for example, the number of units are being increased from nineteen to twenty-four. This will increase the installed capacity from 2,829 mw in 1975 to 6,180 mw in 1985. In Canada 740 mw of hydropower are installed, with an additional 2,940 under construction.

Navigation System

The major navigational waterways of the Columbia system are comprised of: a navigation channel maintained at 12.2 m (40 ft) from the river mouth 175 km (109 mi) to Portland, Oregon; a 4.6 m (15 ft) channel another 72 km (45 mi) to Bonneville Dam; and a 4.3 m (14 ft) channel to Lewiston, Idaho, 692 km (430 mi) above the mouth of the Columbia. Lesser navigational channels are maintained on the Willamette, middle reaches of the Snake, and on other rivers. Seven of the eight dams providing slack water between Bonneville and Lewiston have lock dimensions of 26.2-by-205.7 m (86-by-675 ft). Each of the eight dams has a single-lift lock. Bonneville, which antedates the other dams, presently has a lock 23.2-by-152.5 m (76-by-500 ft). This lock will most probably be enlarged to increase the capacity of the navigation system.

Recreation Resources

Excluding the Canadian drainage, there are approximately one million ha (2.5 million ac) of slack water for recreation, about one-half of which are formed by more than 160 impoundments each containing 6.2 million m^3 (5,000 ac ft) or more of storage. Smaller impoundments for irrigation and livestock watering purposes are not included in this total. About 16,340 km (10,150 mi) of free flowing rivers are classified as either existing and/or potential wild, scenic, and recreational rivers. Federal and state agencies operate many wildlife refuges and facilities for fish propagation. The Canadian portion of the Columbia drainage has a far greater per-capita supply of recreational resources than the U.S. Pacific Northwest.

Waste Disposal and Treatment

Municipal plants provide treatment for practically all human wastes in urbanized areas and for some industrial wastes too. Secondary treatment is becoming prevalent. In the U.S. part of the study region, municipal and industrial waste production account for about 13 and 87 percent of the regional total, respectively, measured by BOD. Irrigation return flows, which often contain significant quantities of salts, nutrients, and pesticides, are estimated to be 2.47 x 10^{10} m^3 (2 x 10^6 ac ft) yearly.

Flood Control and Land Drainage

In the main stem and major tributaries of the Columbia 5.37 x 10^{10} m³ (43.6 million ac ft) of storage are designated for flood control. This would reduce the record peak flow on the lower Columbia by 45 percent. Unregulated flood flows cause considerable damage on some tributaries of the Columbia and on coastal streams of Washington and Oregon. Floodplain regulation, including land use zoning, subdivision regulations, and building codes, as part of the National Flood Insurance Program will also reduce future susceptibility to flood damage in the many jurisdictions participating in the program in the U.S. part of the region.

Over one million ha (2.5 million ac) of cropland have problems related to excess water. These lands are about equally divided between western Washington and Oregon, where excess precipitation and seepage are the major causes, and irrigated lands east of the Cascades, where canal seepage and poor water management cause drainage problems. Works designed to control bank and channel erosion exceed 3,000 km (1,853 mi). Various types of erosion control measures, such as contour plowing, terracing, are practiced on more than 3.6 million ha (8.9 million ac) of cropland.

Organizational Units of Management

A multitude of actors in both the public and private sectors employ a number of strategies of produce water-derived services. During 1974 U.S. federal expenditure on water resources in the region approximated $453 million, which was probably a very significant part of the total water-related expenditures in the study area.

Three categories of organizational units function in the region: (1) those numerous actors whose interests in and responsibilities for water-derived services are functional; (2) those several actors, five states and British Columbia, whose interests are territorial, and (3) two entities with a regional orientation, that have sought to orchestrate the actions of the others. Over the last four decades federally initiated efforts (in the U.S. part of the region) to form a regional unit of water planning have devolved upon the states' increased responsibilities for water resource planning. The present Pacific Northwest River Basins Commission is a partnership of states and federal agencies under the direction of a chairman appointed by the President. In the Canadian part of the study area, the Province of British Columbia exercises considerably more control over water development than the neighboring states. At the international level, the U.S.–Canadian Treaty incorporates the principle of sharing downstream benefits for flood-loss reduction and hydroelectric generation, while the International Joint Commission presides over discussion of a variety of issues concerning boundary waters.

Impacts of the Program

Soil

In the course of developing 3.1 million ha (7.7 million ac) of irrigated land some waterlogging and salinization have resulted. While most of the affected lands remain productive, some of the areas that were developed over a half a century ago without provision of drainage have become unproductive. When salinization results, the soil is the source, as most surface waters used for irrigation in the region contain only 75–150 ppm of salt. Waterlogging and salinization occur together in much of the semiarid parts of the region; salinization is not a problem in the humid lands west of the Cascades.

The drainage required to maintain the productivity of lands affected by waterlogging and salinization has proved costly. For example, in the Columbia basin project, where over 200,000 ha (500,000 ac) were brought under irrigation over the last three decades, drainage costs are markedly higher than anticipated.

A major gap in knowledge, discussed in a later section, is

how to develop an efficacious method of discouraging irrigators from applying excessive amounts of water. In addition,
there is need for more accurate methods of estimating how
much salt is in the soil and how rapidly it can be removed.
Finally, there is need to automate and generally improve
methods of surface irrigation to a degree similar to the improvements made in sprinkler irrigation over the last decades.
This is significant because surface irrigation will remain in wide
use in the region (and in the world) and because it usually
requires much less energy than sprinklers. Research in the
region has generally addressed physical relationships but has
paid relatively little attention to how technological improvements are adopted at the farm level.[2]

Channel

The region has in general few problems associated with
siltation. A notable exception results from the erosion of the
loess-mantled Palouse Hills in southeastern Washington,
causing turbine scour at some of the generating facilities on the
lower Snake River. The operations of most of the 160 reservoirs in the region are not significantly affected by sedimentation; corrective action will not be required for many more
decades. However, the numerous dams in the region have
modified the cross section, gradient, and transport of materials
in downstream reaches. While the precise consequences of the
modifications are not known, they may prove significant in
parts of the region by raising the costs of procuring sand and
gravel for building materials and, in some instances, by
reducing flood control benefits through accretion.

Major research problems associated with water-resource
development and river channels are the following: the rate of
accretion and its impact on potential flood damages in the
lower Columbia; the effects that dams have on downstream
channels; development of methodologies to measure the sedimentation resulting from changes in land use; and improved
methods of predicting and controlling turbidity in reservoirs.

Research in the region that has helped to close gaps in knowledge include several studies by the U.S. Geological Survey on the transport of radionuclides in the Columbia River;[3] and a study of the changing nature and distribution of sediments in the estuary of the Columbia River.[4]

Water Quality

It is difficult to generalize about water quality in the region. Some aspects of water quality in the region are improving despite enlarged population and industrial production. This results because a growing percent of the regional population is served by municipal sewage plants with increasingly higher levels of treatment; and, particularly because more industries are practicing chemical recovery of what was formerly discharged as waste while providing better treatment before discharging the remaining wastes. The Environmental Protection Agency (EPA) reported that during the ten-year period from 1963 to 1972 the BOD discharge in Idaho, Oregon, and Washington decreased by approximately 44, 70, and 59 percent, respectively. Almost all of the decrease in the discharge of the organic pollutants was credited to improved operations of industry.

On the other hand there is an apparent widespread increase in the presence of organic and inorganic toxins from nonpoint sources. The EPA notes, nowever, that the significance of this trend is obscured because there was little sampling of such materials in the past. In addition, downstream from densely populated urban areas there are significant problems of water quality, including excessive counts of coliform bacteria and levels of dissolved oxygen. Further, gas supersaturation and water temperatures that sometimes exceed 20°C (68°F) are serious problems for anadromous fisheries. Supersaturation is particularly serious in reaches of the lower Snake River, while concern is widely expressed about temperatures in the lower Snake and Yakima systems.

Principal research needs include: how to measure the effects of nonpoint sources of pollution, e.g., irrigation return flows and runoff from areas of intensive silviculture; development of methodologies to establish minimum flow requirements for various in-stream uses of water such as fish production and recreation; establishment of the relationships between surface uses of water and ground water quality; and how best to meet the requirements of federal legislation on water quality through changes in production processes, techniques of treatment, and methods of disposal. Although some light has been thrown on these questions by numerous studies, comprehensive approaches at the necessary scale have not been adopted.

Aquatic Life

Water developments in the region have affected aquatic life markedly. Until recently, inadequate consideration of biological factors during the design and/or operation of the projects was often the cause. Over the last several years increasing consideration has been given to aquatic life when planning water resource developments but lack of knowledge about the interrelationships between development and aquatic life is a significant impediment to achieving comprehensive water management even when serious efforts are made to take account of it.

The growth of aquatic plants has not generally created serious problems; in some reservoirs their presence enhances the habitat of waterfowl and warm water fisheries, while impeding recreational use. Aquatic plants in irrigation delivery systems are controlled principally by herbicides, which is sometimes costly. Animal communities (particularly fur bearers) have, in general, been adversely affected by water development. However, at least one large irrigation project (the Columbia Basin Project) has enhanced waterfowl production by creating lakes and marshes. Anadromous fisheries have been adversely impacted; but it is estimated that the annual average harvestable yield of salmon (*Oncorhynchus* spp.), which had declined sharply, has now been increased over the low point by extensive propagation programs. These programs are less successful on the middle and upper reaches of the

Columbia system. For example, salmon runs in the Snake River are jeopardized in the wake of four dams completed on the lower Snake over the last decade. A major problem that has arisen this year is how the states can manage the anadromous fisheries of the Columbia River and adjacent coastal waters in light of a recent interpretation by federal judges of treaties between the U.S. government and various Indian tribes in the 1850s, which antedate statehood. The interpretation stipulates that certain Indian tribes are entitled to harvest one-half of the salmon and steelhead destined to pass over Bonneville Dam during their upstream migration.

More research is needed to assess the complex interrelationships between water development and aquatic life. Although pioneer research on these questions has been carried out in the region over the last several years,[5] plans to use hydroelectric plants increasingly to produce peak power makes it important to intensify research in that direction.[6]

Groundwater

The region has a large reserve of groundwater, which has already become the base for significant economic development in several subregions. Approximately 25 percent of the region is underlain by aquifers capable of yielding moderate to large volumes of water. These aquifers are generally found in basalts of the Snake River group, in basalts of the Columbia River group, and in alluvial deposits. Aquifers in the Snake River group have the greatest sustained yield. Because aquifers in the Columbia River group combine moderate to high productivity with low total storage capacity, they are very sensitive to overdraft.

If present use trends continue, it is estimated that overdraft, degradation of groundwater quality, and waterlogging—singly or in combination—may well take place in several subregions. The adverse effects of overdraft are present or potentially present in parts of Idaho's Snake River plain, eastern Washington and northeastern and coastal Oregon, while waterlogging is present in the eastern end of the Snake

River plain.

Detailed information is needed about deep aquifers and about subsurface conditions in remote, sparsely populated areas of the region. Needed also are inexpensive methods of gathering such data. In addition, a better understanding is required about the broad range of human factors that interact with groundwater resources, including institutional and behavioral aspects that affect the adoption of innovative groundwater laws, integrating the use of surface and groundwaters, transfer of pumping rights, pump taxes, etc.

Productive research on groundwater over the last thirty years entailed comprehensive collection and interpretation of data on the geologic setting and well-water data. Based on the geohydrological appraisal, productive techniques were developed to guide the management of groundwater systems. These techniques are electrical analog models and computerized mathematical models of groundwater flow systems. These models have not been widely implemented, however. As of July 1975, there were less than a dozen such models in the region.[7]

Water Flow

Water flow in the region has been altered in major ways by the 160 reservoirs. Much of the storage is multipurpose but practically all is used to reduce flood flows when necessary. In June 1974, the flood crest which would have occurred on the lower Columbia was reduced three m (9.5 ft) at Vancouver, Washington, preventing an estimated $240 million of flood damages.

Storage also augments low flows, increasing benefits derived from hydropower plants, irrigation, fish migration, recreation, and waste carriage. The positive results of low flow augmentation are mostly widely acclaimed in the densely settled Willamette Valley, where storage releases during the summer double the normal low flow of the river.This was coupled with more advanced and inclusive treatment of wastes, thereby encouraging extensive water-contact recreation after many years of little use due to poor water quality.

The major gap in knowledge is the absence of sound methodologies to assess the impacts of incremental changes in flow on in-stream uses. Some research on this broad question has been completed in the region[8] and a number of research efforts on identification of acceptable minimum flows are underway.[9] In addition, continued efforts by federal agencies to implement the requirements of multi-objective planning will improve methods of determining trade-offs between uses and regions that result from incremental changes in flow.[10] The entire question of water flows is further complicated by uncertainty over the validity of water rights issued by the states, including the results of future adjudications as to how much of the surface waters originating on federal lands might be reserved for federal agencies.

Flows also must be considered in relation to proposals to divert water from the Columbia system to the Pacific Southwest. Prior to the moratorium on diversion studies in 1968, numerous preliminary studies proposed diverting from less than $2.5 \times 10^9 \text{m}^3$ to more than $1.9 \times 10^{10} \text{m}^3$ (2×10^6 to 1.5×10^7 ac ft) annually from the Columbia system. Even though the moratorium will not expire until 1978, the diversion issue is now receiving increasing attention.

Human Health

Water resource development has altered the distribution of vectors in the region, although serious health problems have not become widespread. In several irrigated subregions encephalitis is of concern; waterfowl attracted by irrigation projects winter over, acting as a reservoir for encephalitic viruses. Where health problems associated with projects exist, they usually reflect either inadequate design and/or operational and maintenance procedures. A study probing the relationship between vectors and irrigation projects in central Oregon concluded that ". . . projects embodying all of the facilities for storage, conveyance, utilization, and drainage can be developed without creating major mosquito sources."[11]

More research is needed on the full range of minimizing vectors associated with water developments and documenting the full costs of not designing vector control into projects. The Tennessee Valley Authority has carried out productive research on vector control for decades;[12] more recently, research efforts across the United States have begun to illuminate linkages between water developments and human health.[13]

Electric Power Production

Electric power production has had several clear impacts. Most of the mainstream of the Columbia River and many of its tributaries are now slack water; the region produces far more hydroelectric energy than any other in the United States; potentially high energy costs stemming from a regional lack of fossil fuels have been mitigated by partial substitution of inexpensive hydroelectric energy; the regional per capita consumption of electrical energy for residential use surpasses that of all other regions, being approximately double the national average; and a disproportionate share of industries oriented towards electrical energy supply are represented in the regional mix.

Much less, however, is known about the secondary impacts of electric power production. A significant gap in knowledge relates to the relationships between the massive development of hydroelectric power and the well-being of the region's inhabitants. Until recently it was widely assumed that extensive development of hydroelectric power would yield large benefits to the general public and cause few if any drawbacks, but, increasingly, public interest groups are questioning the validity of regional benefits purportedly stemming from aluminum reduction plants and other industrial consumers of large volumes of low-priced electrical energy.

A primary question today is how to establish priorities for trade-offs that will result from the increasing use of hydroelectric plants to supply peaking power. Research is being accelerated on the manifold impacts of proposed peaking operations on anadromous fisheries, riparian wildlife, recreation, navigation, and other water related phenomena.[14]

Waterborne Traffic

Water resource development has encouraged rapid growth of waterborne traffic on extensive reaches of the Columbia system. Between 1930 and 1974 the volume of goods passing the Bonneville Locks grew from approximately 75,000 to 4 million MT. Excluding maritime and coastal traffic in the lower reaches, the flow of commodities is characterized by downstream movements of wheat and upstream movements of petroleum and petroleum products. A diversification of products is projected to take place over the next ten to fifteen years; significant volumes of agricultural chemicals will move upstream, while more wood products will move downstream. The volume of commodities in coastal and maritime trade handled by ports on the lower 197 km (110 mi) of the Columbia-Willamette system exceeds that carried on other reaches of the river by about eight fold.

Research is needed to identify the impact of inland water transport on patterns of land use, and of proposed lockage and/or higher fuel taxes on the viability of inland navigation. In the region, relatively little research has been done on the first need[15] and none completed on the second.[16]

Agricultural Production

Since much of the region is semi-arid, irrigation has increased the production of agricultural products markedly. In 1974, the sale value of crops, livestock, and livestock products from these irrigated lands is estimated to have exceeded $1 billion.

There are three types of research needs. The first concerns more precise knowledge about crop response to irrigation practices, including drainage requirements for crops. Agronomics and related sciences have made substantial contributions to this end.[17] The second gap in knowledge concerns means of increasing efficiencies in the irrigation process, including: how to educate and persuade irrigators to exercise the best known irrigation techniques; what effects the pricing of irrigation water has on irrigation practices; how to modify water laws that now encourage irrigators to overirrigate as a means of retaining their water rights; and how to provide reasonably secure water rights for efficient irrigators while adding needed flexibility to the systems of water rights in the region. Recent research efforts have developed data and methodologies upon which further studies may be based.[18] The third research need concerns the secondary impacts of irrigation on the redistribution of income and socioeconomic activities, and how large subsidies for federal irrigation projects affect this redistribution. Although some of the facets of this overriding question have been illuminated,[19] much additional research is needed.

Industrial Production

Water developments have indirectly affected industrial production in the region. Large blocks of inexpensive electrical energy have attacted energy-oriented industries such as aluminum reduction plants while large irrigation projects have drawn firms processing potatoes, sugar beets, and a variety of other crops. More information is needed about: the interrelationships between industrial and competing uses of water; how water quality standards affect the location of industry; and the precise withdrawal of water and range of water requirements by industrial users. In the region, little research concerning these questions has been undertaken; moreover, the hypotheses on industrial water use in other parts of the country[20] have not been tested.

Recreation

Over the last three decades the increase of recreational water use in the region has surpassed the growth rate of most other

socioeconomic indices. This increase probably reflects changing socioeconomic conditions more than the doubling of the
area of slack water by the new reservoirs. Some of the many
gaps of knowledge include: how to establish the impact of
incremental flow changes on recreational use; how much
recreation use reservoirs actually stimulate; the effect that
lockage fees and/or increased fuel taxes will have on the sale
and operation of boats and associated recreational equipment;
the trade-offs between slack water and whitewater recreation;
and whether flows below dams can be guaranteed for recreation under the present institutional arrangements. Until recently, relatively little research on recreation aspects of water was
undertaken in the region.[21] Presently, several of the gaps are
being addressed by the Bureau of Outdoor Recreation.[22]

Urban Settlement Patterns

With one exception, the development of water resources has
had a relatively minor role in the spacing and size of urban
settlements. Irrigation of formerly non-productive agriculture
areas has played a major role in several of the semiarid
subregions east of the Cascade Range. In this connection an
early study sought to assess the future requirements of establishing and maintaining towns in and adjacent to the then
unconstructed Columbia Basin Irrigation Project.[23]
The impact of water resource development on internal
patterns of urban occupance is more significant. The form and
direction of occupance are impacted by the provision and
modification of municipal sewer and/or water lines, and by
flood flows. More research is needed on the relationships
between internal patterns of urban occupance and water
development. Findings[24] on these relationships in other parts
of the country have not been tested in the region.

Summary

A variety of strategies have been employed in the Columbia

basin to develop a full range of water-derived services, although flow uses are predominant. Much of the water resource development was accomplished by modifying the spatial and temporal distribution of water. Many earlier projects that provided one or more of the additional water-derived services—electrical energy, flood-flow modification, irrigation, navigation capacity, and municipal and industrial supplies—were often accomplished at the expense of fish and wildlife habitat, water quality, and other factors associated with environmental quality. Over the last two decades a more balanced approach to water resource management has been evolving, reflecting the increased value of leisure time uses of water and heightened sensitivity to quality of the environment. This new approach also places a growing emphasis on the wise management of water use, as opposed to a longstanding preoccupation with continually increasing water supply.

While real and impending conflicts between water users have spurred research efforts over the last decade, more attention should be given to the secondary impacts of water development. Particularly important is the need to develop methodologies to assess trade-offs resulting from incremental changes in stream flow. This will become even more significant as the Columbia system is used increasingly to supply peak power.

Notes

1. G. F. White, *Strategies of American Water Management* (Ann Arbor: University of Michigan Press, 1969).

2. For example, see B. L. McNeal and W. A. Starr, "Potential Salinity Hazards Upon Irrigation Development in the Horse Heaven Hills" (Pullman: Washington Agricultural Experiment Station, February 1974).

3. For an example of several studies on this topic, see J. L. Nelson and W. L. Haushield, "Accumulation of Radionuclides in Bed Sediments of the Columbia River between the Hanform Reactors and McNary Dam," in American Geophysical Union, *Water Resources Research,* Vol. 6, No. 1, pp. 130-137.

4. J. B. Lockett, "Sediment Transport and Diffusion—

Columbia River Estuary and Entrance," in American Society of Civil Engineers, *Journal of Waters and Harbors Division*, Vol 93, No. WW4, Proceeding Page 5601, pp. 167-175.

5. For example, see W. H. Oliver and D. C. Barnett, *Wildlife Studies in the Wells Hydroelectric Project Area,* F.P.C. License No. 2149 (State of Washington, Department of Game, July 1966); P. A. Johnsgard, "Effects of Water Fluctuation and Vegetative Changes on Bird Populations, Especially Waterfowl," *Ecology, Supersaturation Caused by Dams on Salmon and Steelhead Trout of the Snake and Columbia Rivers, 1974* (Seattle, Wash.: Northwest Fisheries Center, National Marine Fisheries Service).

6. A number of research efforts sponsored by the Corps of Engineers' Columbia River and Tributaries Review Study (CRT). The CRT is to determine if any changes or additions should be made in the Corps of Engineers' and other projects on the Columbia System. An example of CRT activities was sponsorship of a six month study by University of Idaho researchers to assess the impact on fish of peaking operations at dams on the lower Snake. Completion date of the research was scheduled for February 1976. Another example of CRT activity is the 429 page *Environmental Assessment Manual* prepared for the Corps of Engineers, North Pacific Division, by Battelle's Pacific Northwest Laboratories. This manual was printed in May 1974 and distributed by the Corps' North Pacific Division.

7. Letter from B. L. Foxworthy, U.S. Geological Survey, July 17, 1975. For problems related to use of basalt aquifers in the region, see H. R. Doerksen, *Regional Problem Analysis in the Pacific Northwest* (Pullman, Wash., Idaho Water Resources Research Institute, Oregon Water Resources Research Institute, and State of Washington Water Research Center, March 1975), pp. 69-102.

8. For an assessment of past and planned research on this topic, see Ad Hoc Instream Flow Study Evaluation Committee, Pacific Northwest River Basins Commission, "Evaluation Report: Need for a Coordinated and Expanded Program of Instream Flow Evaluation Data in the Pacific Northwest" (unpublished report, August 1974). See also Pacific Northwest

River Basins Commission, *Anatomy of a River: An Evaluation of Water Requirements for the Hell's Canyon Reach of the Snake River, Conducted March, 1973* (Vancouver, Wash., July 1974).

9. For example, under the auspices of the Corps of Engineers' Fisheries Engineering Research Program, the Fish Commission of Oregon is determining the impact of power plant discharge rates on fish collection efficiency, while the State of Washington's Department of Fisheries is researching the effect of peaking operations on the stranding of juvenile salmonids.

10. *U.S. Federal Register,* Vol. 38, No. 174, Part III, September 10, 1973, pp. 24778-24869.

11. Aquatic Plant and Insect Control Committee, *Report on a Study of Mosquito Problems Associated with Development of the Crooked River Irrigation Project, Central Oregon, 1960-1966,* Pacific Northwest River Basins Commission and U.S. Water Resources Council, October 29, 1968, p. 23.

12. For example, see O. M. Derryberry, "Malaria Control on Water Resource Development Projects: The 25 Year Experience of TVA," *Proceedings of the Sixth International Congress on Tropical Medicine and Malaria,* Vol. 3; and *Environmental Statement: Vector Control Program,* TVA-EP-EIS-74-2 (Chattanooga, Tenn., April 1974).

13. Institute of Rural Environmental Health–Colorado State University, *A Study of Mosquito Prevention and Control Problems Associated with Stream Modification Projects* (Ft. Collins: Colorado State University, July 1974); and U.S. Department of Health, Education, and Welfare/Public Health Service, *Prevention end Control of Vector Problems Associated with Water Resources* (Ft. Collins: Colorado State University, 1965, reprinted 1975).

14. Under the CRT Program several studies will be undertaken, including Impacts of Peaking on Riparian Wildlife; Bypass of Juvenile Salmonids Around Turbines; and Effects of Peaking on Juvenile Salmonids. For an earlier study that addresses problems related to peaking, see U.S. Army Engineer District, Portland, Oregon, *Final Environmental Statement:* Modification for Peaking the Dallas to Vancouver—

Columbia River, Oregon and Washington (Portland, Oreg., February 1972).

15. G. Torget, "Upper Columbia Navigation Extension Report" (unpublished study).

16. An ongoing study by the Corps of Engineers' Seattle and Portland District Offices entitled "Bonneville Lock Replacement Study" (February 1976), reportedly gives some attention to this aspect.

17. For example, see R. M. Hagan, H. R. Haise, and T. C. Edminster (eds.), *Irrigation of Agricultural Lands* (Madison, Wis.: American Society of Agronomy, 1967).

18. Washington State University, College of Agriculture, *Horse Heaven Hills Irrigation and Development Potential* (Pullman, Wash. 1970). Two research projects in the Department of Agricultural and Resource Economics at Oregon State University also address some of these questions: "Oregon's North Columbia Basin: Irrigation System Development Project" (August 1976); and "The Economics of Water Pricing: Effects of Alternative Prices of Water on Farm Income," scheduled for completion in September 1976. Increased flexibility would result if term permits for water were issued rather than traditional appropriative water rights. This is being considered by the state of Washington.

19. Five of many examples are: J. T. McBroon, *Towns and Villages,* Columbia Basin Joint Investigations, Columbia Basin Project, Washington, Problem 18 (Boise, Idaho: Bureau of Reclamation, June 1947); M. E. Marts, *An Experiment in the Measurement of Indirect Benefits of Irrigation—Payette, Idaho* (Boise, Idaho: Bureau of Reclamation, June 1950); N. D. Kimball and E. N. Castle, *Secondary Benefits and Irrigation Project Planning* (Corvallis, Oreg.: Agricultural Experiment Station, Technical Bulletin 69, May 1962); C. L. Infanger, *Income Distributional Consequences of Publically Provided Irrigation: The Columbia Basin Project* (Ph.D. Dissertation, Department of Agricultural Economics, Washington State University, 1974, Published by the National Technical Information Service); Washington State University, College of Agriculture, *Horse Heaven Hills Irrigation.*

20. Notable, because it refuted the concept that a large

supply of good quality water was one of the most significant factors in the location of industry, is: G. F. White, "Industrial Water Use—A Review," *The Geographical Review,* Vol. 50, No. 3 (July 1960), pp. 412-430. Also see C. W. Howe, "Water Resources and Regional Economic Growth in the United States, 1950-1960," *The Southern Economic Journal,* Vol. 34, No. 4, April 1968, pp. 477-489.

21. For a method used by federal agencies to estimate recreational use of future reservoirs, see *Estimating Initial Reservoir Recreation Use* (Plan Formulation and Evaluation Studies—Recreation, Technical Report No. 2, October 1969). For an account of institutional problems related to intergovernmental cost sharing of recreation at federal reservoirs, see K. W. Muckleston, *The Problems and Issues of Implementing the Federal Water Project Recreation Act in the Pacific Northwest* (Corvallis, Oreg.: Water Resources Research Institute, WRRI-20, October 1973). For a summary of regional problems of analyses regarding wild and scenic rivers, see Doerksen, *Regional Problem Analysis,* pp. 89-122.

22. Under the CRT Program the U.S. Bureau of Outdoor Recreation (BOR) and state agencies responsible for recreation will address some of these questions. For example, the BOR is budgeting six man months to an ongoing study entitled "Irrigation Depletion/Instream Flow Study." For questions related to recreational use of wild and scenic rivers in the region, see Doerksen, *Regional Problem Analyses,* Part C, pp. 86-122.

23. McBroom, *Towns and Villages.*

24. Several research papers published by the department of Geography at the University of Chicago addressed various facets of this question. See, for example, G. F. White et al., *Changes in Urban Occupance of Flood Plains in the United States* (Chicago: The University of Chicago, Department of Geography Research Paper No. 57, 1968). Also see Rivkin/ Carson, Inc., *Population Growth in Communities in Relation to Water Resources Policy,* PB 205 248 (Springfield, Va.: National Technical Information Service, October 1971), especially pp. 114-150.

6

Changing Approaches to Water Management in the Fraser River Basin

W. R. Derrick Sewell

An important characteristic of water management institutions is their tendency towards inertia. Laws, administrative structures, or policies may remain unaltered for decades, even though the need for modification has been frequently pointed out. Basic changes, it seems, often come only as a result of a crisis, such as a severe physical event (like a flood or a drought), an intractable conflict between two or more resource uses, or a convincing challenge to long-held concepts or social values.[1] There are numerous examples in North America and elsewhere.[2] The tragedy is that reliance on stress as a signal for action may be very costly in economic, social, or environmental terms. Moreover, it may not produce a basic change in approach but rather an intensification of the existing one. At the same time, however, crises can play a very valuable role. In arousing attention to a particular problem, for example, it may create better understanding of the issue, and it may also generate the public support needed for implementation of corrective measures.

W. R. Derrick Sewell is professor of geography, University of Victoria, Victoria, British Columbia. The author wishes to acknowledge the assistance of Christianna Crook who undertook library research and who contributed many useful ideas, Lorna Barr who provided helpful comments on an earlier draft, and Ben Marr, B.C. Deputy Minister of Environment, who furnished background information and offered many useful suggestions.

Experience in managing the water resources of the Fraser River Basin in British Columbia provides an interesting illustration of the role of stress in shaping water planning and policymaking. It shows not only that there has been a tendency to rely on crisis to identify problems but also that planners and policymakers seem to have depended on such events to furnish support for recommended solutions. In addition, the Fraser River experience seems to suggest that different kinds of problems require different kinds of stress to stimulate action. In the case of floods, for example, physical events have generally provided the basic stimulus, while in the case of water pollution or of salmon preservation, resource conflicts or threats to valued elements of the environment seem to have been the main goads to action.

The Problems and Their Setting

The Fraser River is one of North America's largest and most important rivers (Figure 6-1). It drains some 90,000 mi^2, an area roughly equivalent to that of West Germany, Romania, or Colorado. It flows 850 mi from its headwaters in the Rocky Mountains to its mouth on the Pacific Ocean. Its average discharge at Hope (some 100 miles from its mouth) is about 92,000 ft^3/s, placing the Fraser as sixth among North America's major rivers.[3]

The Fraser River has a snow-fed regime, featuring low flows during the winter months and high flows in the spring. There is a considerable range between these flows. The mean monthly discharge at Hope in March, for example, is 28,600 ft^3/s, whereas the mean monthly flow at the same point in June is 251,000 ft^3/s (Figure 6-2). The larger part of the streamflow originates in the eastern mountain region which flanks the basin. This region constitutes about 36 percent of the drainage area upstream from Hope but it contributes some 60 percent of the runoff. The interior plateau constitutes about 62 percent of the drainage area but contributes only 36 percent of the runoff. The remaining 4 percent of the runoff comes from the Coast Mountains in the southwest corner of the catchment area.[4]

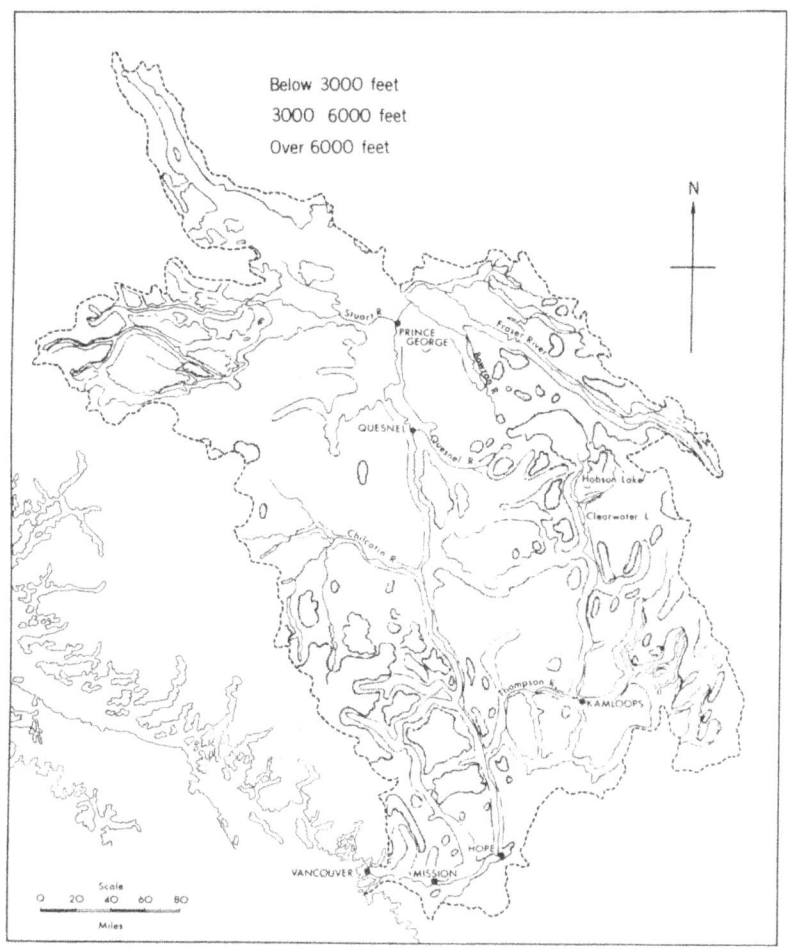

Below 3000 feet
3000 6000 feet
Over 6000 feet

FIG. 6–1 FRASER RIVER BASIN PHYSIOGRAPHY

The basin possesses vast resources of forests and minerals. Together these provide the basis for large and expanding primary and secondary industries, particularly in the central and northerly portions of the basin. There are major sawmills, for example, in and around Prince George, Smithers, Quesnel, Williams Lake, and Kamloops. The Prince George District, for example, has a daily sawmill capacity of some 6,148 FBM.[5] Although much of the province's pulp and paper manufacture takes place at the coast, there are also mills in the interior of the basin, notably at Prince George, Quesnel, Mackenzie, and

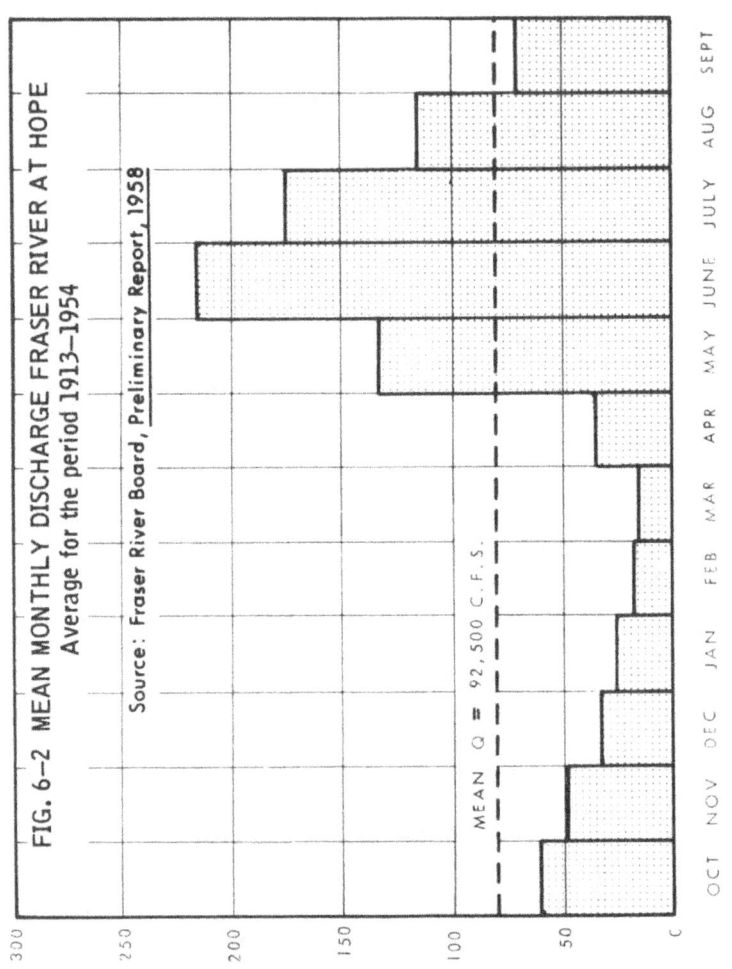

FIG. 6–2 MEAN MONTHLY DISCHARGE FRASER RIVER AT HOPE
Average for the period 1913–1954

Source: Fraser River Board, Preliminary Report, 1958

MEAN Q = 92,500 C.F.S.

ANNUAL MEAN MONTHLY DISCHARGE IN '000 C.F.S.

MONTHS OF THE WATER YEAR

OCT NOV DEC JAN FEB MAR APR MAY JUNE JULY AUG SEPT

Kamloops. Large copper mines, with capacities of over 5,000 tons per day are located at Babine Lake. Major molybdenum mines are situated at Endako, and Gibralter (near Williams Lake).[6] There are large range lands in the interior, supporting a thriving beef industry. Rich alluvial soils have furnished the basis for dairy farming and market gardening industries in the lower reaches of the river. In addition, the Fraser supports one of the world's most valuable salmon resources.[7] Each year about 10 million salmon migrate up the river to spawn and reproduce. Some two to four years later more than 300 million young fingerlings descend to the Pacific Ocean to spend the next portion of their lives. These salmon not only provide a livelihood for some 10,000 British Columbian fishermen and perhaps another 4,000 shoreworkers, but also are a source of food and income for native Indian populations,[8] and a focus for a rapidly expanding sports fishery.[9] They are, in addition, the basis for fishing industries of several other nations, including the United States, Japan, and the Soviet Union. The annual catch in British Columbia alone is valued at $45 million.

Today some 1.5 million people make their home in the Fraser River basin. About 20 pecent of them live in the interior of the basin, above Hope. There are several small towns in the region, including Prince George, Quesnel, Williams Lake, and Kamloops, each of which is a service center for numerous smaller communities engaged in farming or various primary production activities. The remaining 80 percent of the basin's population is concentrated into a tiny, wedge-shaped area stretching from Hope to the sea, a distance of about 100 miles.

The picture which emerges is one of a very large river, flowing through a rich storehouse of natural resources, which thus far has been developed to only a limited extent. The same is true of the water resource itself. It could be developed to harness a hydroelectric power potential estimated at some 8 million kw,[10] but so far only a few dams on some of the smaller tributaries have been built in the basin, with a total capacity of some 750,000 kw.[11] A diversion of the Nechako River (a tributary in the northeastern part of the basin) to the Kemano River on the Pacific coast has made possible the development of a 900,000 kw power project to serve the Kitimat aluminum plant. (This diversion, incidentally, is sometimes used as a means of controlling flood flows on the Fraser River.) There are also opportunities for irrigation in some parts of the basin. In total, about 200,000 ac of land could be irrigated in the basin.[12] Presently, however, only half of it has been developed, requiring some 194,600 ac ft of water. Most of this development is in the Kamloops-Lillooet area in the central part of the basin.

The Fraser River is navigable from its mouth to Hope, and along sections of the river elsewhere in the basin.[13] Deep-sea ships can navigate from the mouth to Port Mann, a distance of 22 mi from the mouth. Over 30 ft of water can be found in that stretch by making use of tides, throughout most of the year. New Westminister, on the north bank, receives more than 500 deep-sea vessels and in excess of 3,000 coastal vessels each year. Tonnages of cargo involved exceed 1 million tons and 4 million tons, respectively. The Fraser River and its tributaries are also used extensively for log booming and barge traffic, and for pleasure boating. The federal government undertakes a continuous dredging program, particularly in the stretch from Mission to Port Mann. Dredging requirements increase considerably following the spring freshet, and are especially large when the freshet is high.

The basin has also become the focus for outdoor recreation, particularly where water-based sports and wilderness experiences are involved. The basin contains many large lakes and several of these are extensively used for recreation. In addition, the salmon sport fishery at the coast, and trout fisheries inland have become major attractions for the local residents as well as tourists.[14]

The water resources of the basin are ample, and most needs can be satisfied without undue difficulty. There are, however, two major problems resulting primarily from the concentration of population and industry in the lower Fraser Valley. Action in dealing with them has tended to be slow and sporadic, and in both instances crisis has played an important role in planning and policymaking.

The Flood Problem

The 1948 Flood

In 1948 British Columbia suffered the most disastrous physical event in its history. As Bruce Hutchison describes it:

> Modern Canada has never seen a flood like this. Quietly, slowly, inexorably, the brown water rose beside the farmland below the canyon and the delta at the river's mouth. Before it earth dykes crumbled and dissolved like sugar, or suddenly, under pressure from below, exploded, tossing trees, stumps and barns into the air. A cargo of uprooted trees and flotsam from the interior poured into the sea, and with it poured the carcasses of milk cows, horses, pigs and sheep to be washed up on the rocks of the Gulf Islands, and even beaches around Victoria. The two transcontinental grades were submerged. Except by air, Vancouver was isolated from the rest of Canada.[15]

The fact that there had been a heavy snowpack in the preceding winter, and that runoff had been delayed by a late spring, had given some cause for concern, and a flood warning was issued on May 1, based on snow survey results. The warning was noted by local authorities but there was no general alarm, for such conditions had often happened before without serious results. Suddenly, however, temperatures began to mount, and continued to do so. The Fraser, Thompson, and all the major tributaries began to rise together. By May 25 the flows of the Fraser River at Mission had reached 18.82 ft, just over 1 ft below the danger mark (20 ft on the Mission gauge). The next day the Mission dyke broke, and was followed by other breaks further down the valley. For over thirty days the valley was in a state of siege by the flood waters.

When the waters receded, they left in their train a scene of devastation.[16] Some 55,000 ac of agricultural land had been inundated (representing about one-third of the floodplain area), causing considerable damage to buildings and machinery, ruining crops, and drowning livestock. On some farms, fields were covered by more than 4 ft of silt. Topsoil was

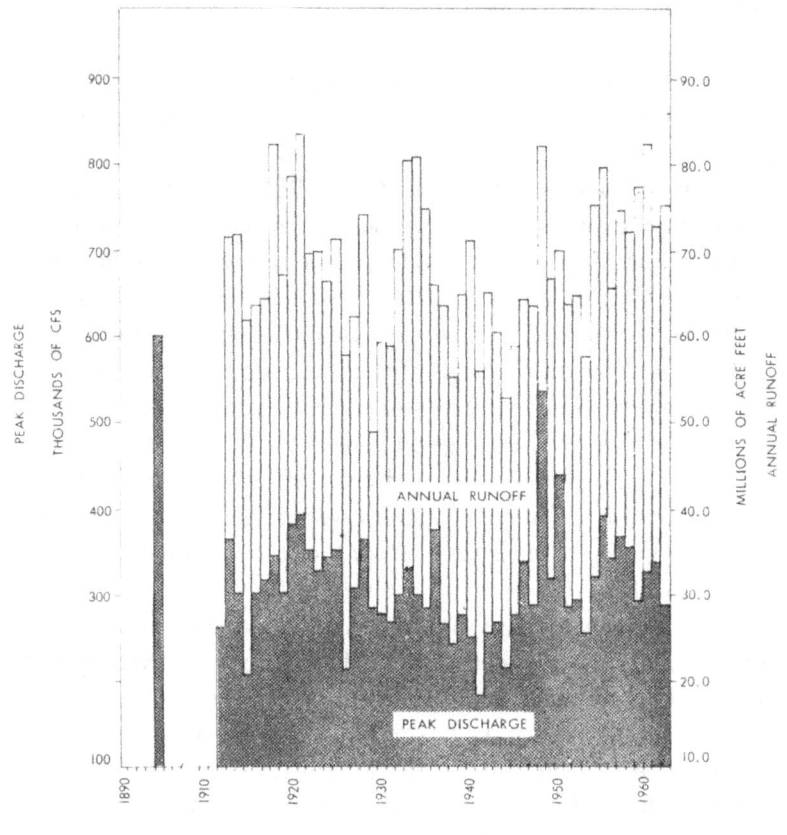

Sources: Fraser River Board, Interim Report, and Department of Northern Affairs and
National Resources, Water Resources Branch data.

FIG. 6–3
FRASER RIVER AT HOPE
PEAK DISCHARGE AND ANNUAL RUNOFF

washed away in many places. Some branches of agriculture in the valley never recovered. More than 2,000 homes were damaged or destroyed, and numerous industrial and commercial establishments suffered damages. There were also major losses beyond the floodplain, particularly as a result of the interruption of communications.

TABLE 6-1

FRASER RIVER AT MISSION RECORDED DATA ON THE

SIX HIGHEST ANNUAL FRESHETS, 1894-1973

	1894	1948	1950	1964	1967	1972
Peak water elevation (ft MSL)	26.0	24.98	24.44	22.96	22.97	23.56
Days above 18 feet	33	38	29	48	34	49
Days from 18 feet to peak*	12	18	8	16	20	24
Days within 0.5 feet of peak*	4	14	6	9	7	4

*Including day on which peak occurred.

SOURCE: Fraser River Joint Advisory Board, <u>Fraser River Upstream Storage Review Report</u>.

No complete estimate of the losses from the 1948 flood is available. Payments by the government for relief and rehabilitation amounted to over $20 million, but actual losses were possibly several times that figure. Flooding occurred in a number of other parts of the basin, notably in Kamloops and Prince George, but effects were minor in comparison with those in the lower Fraser Valley.

The 1948 flood was certainly the most damaging of all the freshets experienced to date, but it was not the biggest flood (Figure 6-3). The 1948 flood reached a peak of 24.98 ft at Mission, whereas the 1894 flood reached a peak of 26.00 ft. The peak flow at Hope in 1894 is estimated at 620,000 ft³/s; that in 1948 reached 536,000 ft³/s.[17] The 1948 flood cannot be considered an especially rare event. Flows approaching that level have occurred on four occasions in the past twenty-five years (Table 6-1).

Studies undertaken by government agencies in British Columbia suggest that during the ninety-three year period 1858–1950 there were at least twenty-four damaging floods, that is, on the average, one every four years.[18] These studies also suggest that a 1948 magnitude flood has a chance of occurrence of about once in sixty years and an 1894 flood about once in 160 years.[19]

Reactions to the Flood

The 1948 flood was a tremendous shock. It gave rise to a call for "firm action" and "innovative policies" to ensure that "such devastation could be avoided in the future."[20] The governments of Canada and of British Columbia moved quickly to deal with the situation. The first action was to deal with the flood victims. A Fraser Valley Rehabilitation Authority was established to administer a $5 million fund provided by the two governments to help those who had lost their homes or suffered severe property losses. The second action was to arrange for the reconstruction of the dykes. A Fraser Valley Dyking Board was set up for this purpose. The federal government agreed to provide some 75 percent of the estimated $11 million cost of repairing the dyking system. The provincial government provided the remainder.[21]

Relief and rehabilitation, dyke reconstruction, and emergency evacuation had long been the main methods of dealing with flood problems in British Columbia. It had become clear, however, that this was insufficient. A call was made for a much more enlightened approach in which floods were viewed as one of several aspects of water management, and in which the lower Fraser Valley was regarded as an integral part of a much larger entity, the Fraser River basin. This had been discussed long before the 1948 flood, but no action was taken. The flood seemed to give further strength to the argument.

The real innovation following the 1948 flood was accomplished by the third action. A succession of engineering boards were established, aiming to provide a much better understanding of the origins of the flood problem and to explore the possibilities and implications of alternative solutions within a basin context. A Dominion-Provincial Board, Fraser River Basin, was established in 1948 to study and report on the water resources and requirements of the area comprising the Fraser

River watershed, with primary attention to power, fisheries, floods, water supply, and recreation. This represented the first attempt at multiple-purpose water resource planning in the basin,[22] and one of the first in Canada. But it suffered from all the problems of first attempts. It was faced with a lack of personnel skilled in comprehensive planning, a lack of data, and a diminishing feeling of urgency as the memory of the flood receded. The board succeeded in encouraging the development of improved programs of data collection but produced no plans for dealing with the flood problem. It was replaced by a new entity in 1955, the Fraser River Board, which was given somewhat more concrete terms of reference and a series of deadlines by which various studies must be completed.

The Fraser River Board produced its final report in 1963,[23] recommending the improvement of the 233 mi dyking system, and the consideration of a program of upstream storage development. The latter would be confined to the upper part of the basin so as to minimize the impact on the salmon fishery. It would also consist of projects that would be economically viable by virtue of the fact that they would produce hydroelectric power at a cost competitive with other sources of power. The recommended scheme would have an average annual output of some 785 mw of firm power and would cost an estimated $403.5 million.

The federal and provincial governments accepted the report, and commended the board on their efforts, but took no immediate action. The recent conclusion of agreements to build the Peace River and Columbia River projects had preempted British Columbia power markets for at least the next twenty years, and residents of the lower Fraser Valley were less than enthusiastic at the prospect of having to make an enlarged contribution to the costs of flood protection. High water was experienced in the spring of 1964 but the dykes held and no major damage was done. The governments diverted their attention to other matters. The apprehension of a major inundation, however, did not die, particularly among the technical advisors. They were cognizant of the fact that a flood of the magnitude of that in 1948 would cause at least $300 million damages in the lower Fraser Valley alone, and could

lead to substantial loss of life. Public interest, however, seemed
to have subsided.

The establishment of a joint Canada–British Columbia
Committee to review the proposals of the 1963 report indicated
a revival of governmental concern. The committee presented a
report in 1967 recommending the implementation of a program of dyke reconstruction, estimated to cost some $33
million, and a detailed review of the program of upstream
storage suggested by the Fraser River Board. This program,
designated System E, is illustrated in Figure 6-4.

The two governments accepted the recommendations and
signed a joint agreement to carry them out. The dyke reconstruction is well underway, and the review of the upstream
storage system has recently been completed.[24] The Fraser
River Joint Advisory Board which undertook the latter study
has recommended the construction of an upstream storage
system, estimated to cost some $1 billion (Table 6-2). It would
consist of seven dams, one of which (lower McGregor) would
be used to divert flood flows from the upper Fraser into the
Peace River watershed. The diversion flows would enter
Williston Lake and would substantially increase the electric
power production at Bennett Dam, and other dams that may
be constructed further downstream on the Peace River (Figure
6-5). This proposal is vigorously opposed by various environmental groups, both in the Peace River area and in southern
British Columbia, on grounds that it would result in considerable disruption of wildlife habitats, and would reduce wilderness recreational opportunities in that area. An alternative
would be to construct a nondiversion dam at the lower McGregor site. It would perform roughly the same flood control
function as the diversion dam, but would result in much less
power production. This would have important implications for
the economic feasibility of the flood control scheme.[25]

A fourth line of action undertaken in the past five years has
been the development of flood mapping programs and land use
regulations. These strategies had been referred to in various
reports, including those of the Fraser River Board, but never
treated very seriously. However, a major flood in 1972 in the
Oak Hills area of Kamloops resulted in huge losses of residen-

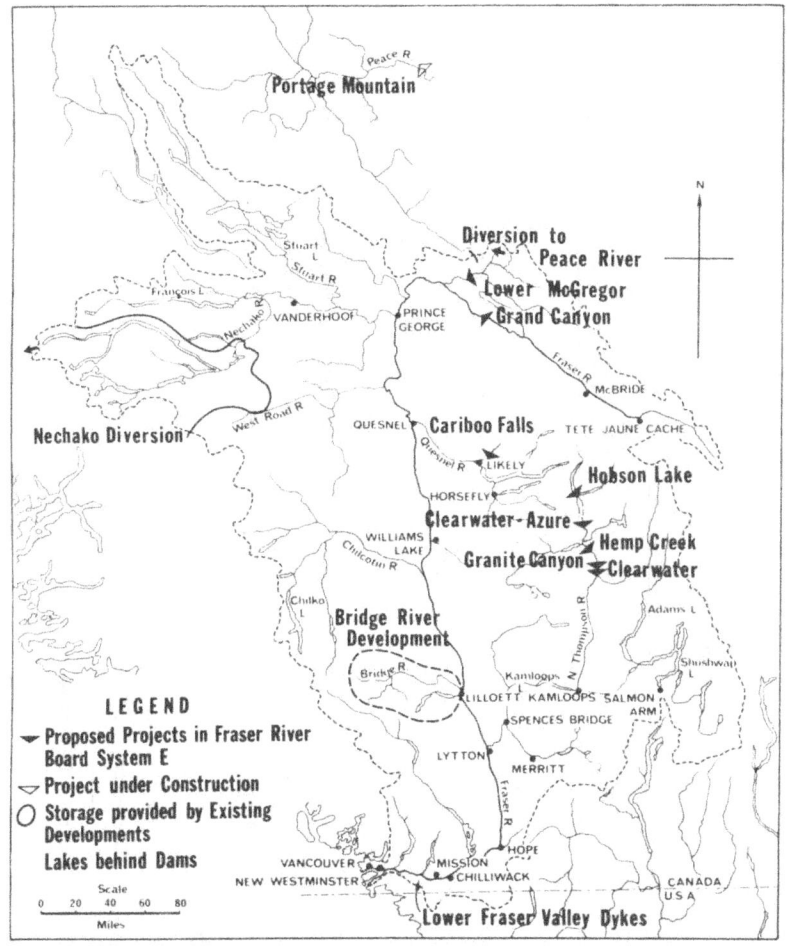

FIG. 6–4
FRASER RIVER BASIN SCHEME OF HYDRO–ELECTRIC
POWER AND FLOOD CONTROL RECOMMENDED BY THE
FRASER RIVER BOARD

tial property, and the provincial government was faced with claims for compensation amounting to more than $12 million. This stimulated the provincial Water Resources Branch to initiate a program of flood mapping, designed to identify much more accurately the areas of the province that would be severely affected by floods. The initial concentration was on the North Thompson River, a major tributary of the Fraser,

TABLE 6-2

SYSTEM E AND ALTERNATIVE PROJECTS

Project	River	Function*	Dam (Height in feet)	Reservoir (Live storage in acre feet)	Generating Capacity (kw)	Estimated Cost** ($millions)
SYSTEM E PROJECTS						
Lower McGregor	McGregor	D	460	---	---	136
Grand Canyon	Fraser	S G	180	1,961,000	144,000	119
Cariboo Falls	Cariboo	S G	240	1,157,000	126,500	56
Hobson Lake	Clearwater	S G	110	811,000	72,000	50
Clearwater-Azure	Clearwater	S G	275	2,263,000	157,000	92
Hemp Creek	Clearwater	S G	450	788,000	368,000	154
Granite Canyon	Clearwater	G	225	Pondage	148,500	62
Clearwater	Clearwater	G	125	Pondage	59,500	29
ALTERNATIVE PROJECTS						
McGregor	McGregor	S G	440	4,260,000	276,000	174
Hobson Lake	Clearwater	S G	122	811,000	72,000	49

* D = Diversion; S = Storage; G = Generation

** Mid-1972 prices.

the Columbia River at Golden, and the Okanagan Valley. Some 140 mi of river, close to settled areas, have been covered so far. Maps will indicate areas subject to inundation by a 200 year flood, generally at a scale of 1 in to 500 ft.

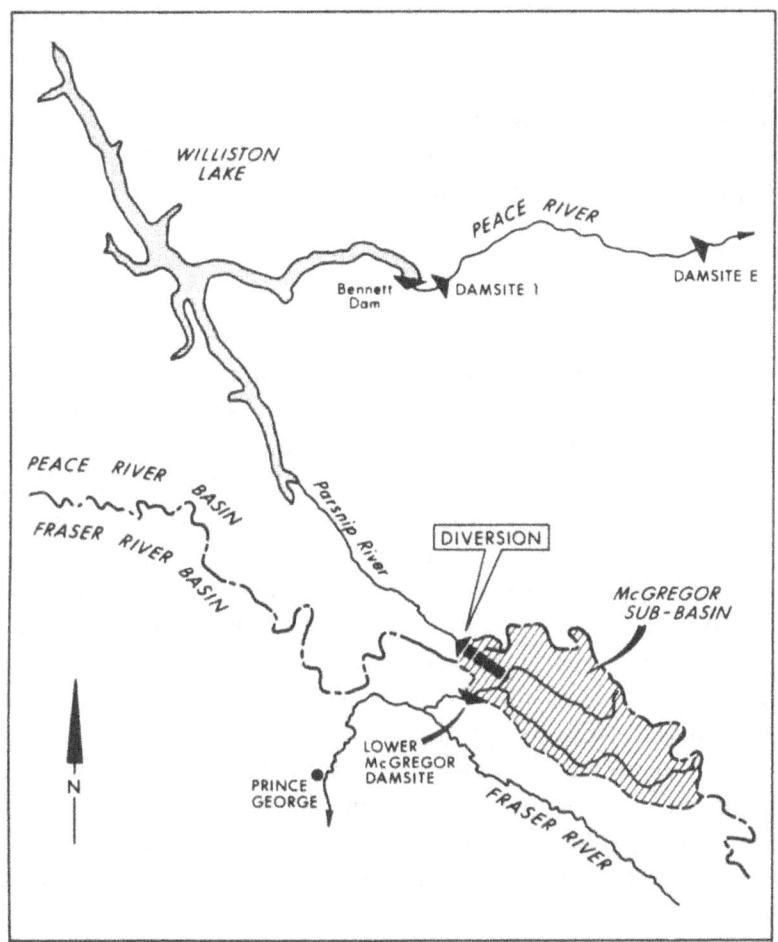

FIG. 6–5 McGREGOR DIVERSION TO PEACE RIVER

Close on the heels of the mapping program has come floodplain zoning. Although such zoning was possible under the 1949 B.C. Town Planning Act, little concrete action was taken. Municipalities, fearing that such regulation might result in an erosion of their tax base, have tended not to enact or implement bylaws which would restrict land use on grounds of potential flooding. A change of government in 1973, however, provided the direction for such action. Shortly after it took office, the government set up a B.C. Land Commission to direct future land use in the province, and particularly to ensure that agricultural land was preserved. Much of this land in the lower Fraser Valley is in the floodplain. Not only will the freeze placed on agricultural land help to prevent further urban encroachment (and hence minimize the flood loss potential), but the passage of the B.C. Land Registry Amendment Act, 1974, will further ensure that new subdivisions will not be built in areas subject to major inundations.[26]

Deteriorating Water Quality

The second major water management problem in the Fraser River basin concerns the deteriorating quality of the river in several parts of the basin. There are several relatively isolated pollution problems in some areas of the interior, notably in areas where pulp and paper industries are located. The major problems, however, are in the lower reaches. The concentration of population and industry in the lower Fraser Valley has led to large discharges of domestic sewage, industrial effluents, and various pesticides and herbicides. Today, the river receives some 214 million gal of treated sewage per day from Vancouver, New Westminister, and various municipalities further upstream.[27] In addition, some eighty-eight industries discharge various kinds of effluent into the river:[28] twelve of them have discharges in excess of 1 million gal per day. Most of these industrial effluents enter the river at points close to its mouth.

Changing Attitudes to Pollution

Until very recently the matter of pollution of the Fraser River was not regarded as a problem of major consequence. As noted earlier, the flow of the river is considerable, even during the winter period. Beyond this there have been no major health epidemics nor fish kills that could be attributed directly to the quality of the river. Also, as pointed out in several recent studies,[29] the major problems of water pollution may pass undetected for a long time because the parameters are not well understood or data on these matters are not normally collected on a systematic basis.

The past decade in British Columbia has witnessed a transformation of attitudes relating to water quality in the lower Fraser from apathy to growing concern. It is not easy to account for this change for there has been no dramatic physical event to draw attention to the problem. Stress in this case seems to have come from questions raised by environmentalists and scientists on the one hand, and by tensions between the federal and provincial governments on the other. The National Conference on Pollution and Our Environment sponsored by the Canadian Council of Resource Ministers in 1961 drew attention to the fact that many aspects of the Canadian environment were experiencing declining quality and that little was being done to correct this situation.[30] There was legislation on the books to deal with such matters but it seemed that it was seldom applied and generally ignored. Situations had been reached in the management of some water bodies, notably the Great Lakes, the Ottawa River, and the St. John River, where further deterioration was expected to have disastrous consequences. The environmental movement took cognizance of these claims and demanded further investigation.[31]

The federal government, perceiving both a need to respond to criticism of its lack of action in the past, and an opportunity to gain prestige by taking a leadership role, passed new legislation, introduced new policies, and undertook a number of reforms in its administrative structure.[32] Perhaps the most important of these developments was the passage of the Canada Water Act, 1970. This act offers cooperation with the provinces to combat water pollution, and, in those instances where provincial action seems inadequate, provides for the federal government to act unilaterally.

The government of British Columbia, unwilling to accept an increase in federal involvement in water pollution control, responded by passing the B.C. Pollution Control Act and by establishing administrative machinery to carry out its provisions. A Pollution Control Board, composed of government-appointed members, and a Pollution Control Branch in the Department of Lands, Forests, and Water Resources were set up. Henceforth, dischargers of effluents in the province would be required to apply for a license for this purpose. The board is empowered to determine what constitutes a polluted condition, to prescribe standards for effluents to be discharged, and to set up technical committees to provide the board with advice. The board issues objectives for industries and municipalities in British Columbia. The branch receives applications for discharge permits and determines whether applications meet the standards set by the board.

Investigating the Problem

One thing was clear in 1970: understanding of the nature and magnitude of the pollution problem was inadequate. Although certain kinds of information had been collected—such as that on BOD and the incidence of pathogens and of various nutrients—little was known about spatial variations in the incidence of particular pollutants and where the latter originated. There was almost no indication of the extent to which various toxic elements were present at various points along the river, nor the sources of them.[33]

In 1974 the B.C. Pollution Control Branch initiated a comprehensive investigation of pollution in the Fraser River watershed, focussing principally upon the mainstream and the major tributaries. This investigation involves studies of some thirty-two chemical and physical parameters. Biological studies are also being undertaken by the branch in the lower Fraser.

Besides the investigations undertaken by the provincial government, the federal Department of Environment has several programs underway. One of these is an inventory of environmental knowledge of the Fraser River estuary which was completed in 1974.[34] It reviews currently available information on the physical and natural aspects of the estuary, and the problems relating to land use and waste disposal. A related investigation focussed on the water resources and land uses of the Strait of Georgia–Puget Sound basin.[35] The department also has several research programs investigating the impacts of various pollutants on fish populations.

The most comprehensive set of investigations, however, has been undertaken by the Westwater Research Centre, located at the University of British Columbia. A four-year program of study was completed recently, funded by research grants from the federal and provincial governments and various independent agencies. These studies have helped place the problem in perspective, and have contributed substantially to understanding particular aspects of it.

Briefly, the Westwater studies conclude that overall, there are no major water quality problems that seriously threaten human health or the survival of fish populations in the lower Fraser. Nevertheless, there are serious problems in some parts of that region, particularly as one moves towards the mouth of the river. The investigations show, for example, that average levels of bacterial pollution rise dramatically between Mission and New Westminister, making bathing and boating progressively less acceptable. Similarly, the presence of fecal coliforms increases rapidly between Hope and the mouth of the river, making withdrawals for drinking purposes unwise (Table 6-3).

The Westwater studies also point out that the potentially most dangerous pollutants—toxic elements—have received little attention so far. From the empirical evidence presented in their reports, it is apparent that while the overall problem of toxicity is not yet serious, some trace elements have already reached critical levels, notably zinc and lead.[36] Further expansion of electroprocess industries and failure to remove trace elements in treatment of domestic sewage could result in other trace elements also reaching unacceptable levels.

The Need for Institutional Change

Much has been accomplished as a result of administrative reform, legislation, and policy changes in controlling water

TABLE 6-3

FECAL COLIFORMS IN THE LOWER FRASER

Station	Fecal Coliforms	Acceptable Water Use
Hope	130	Borderline drinking
Mission	775	No drinking or swimming
Patullo Bridge	2750	No use
Annacis Island	1130	Marginally acceptable for non contact recreation and irrigation
Garry Point	880	Same as Annacis Island
Queensboro Bridge	7730	No use
Fraser Street	16480	No use
Oak Street	5110	No use
Dunsmuir Bridge	2550	No use

SOURCE: A. Dorcey (ed.), The Uncertain Future of the Lower Fraser, p. 26.

pollution in the lower Fraser. There are indications that the governments are committed to action; and there is a vigorous research effort to identify the dimensions of the problem and to explore the possibilities and implications of alternative ways of dealing with it.

It has also become clear, however, that further institutional changes may be required if the problem of water quality management on the lower Fraser is to be solved. Three matters in particular need urgent attention. The first concerns the lack of information on such matters as nonpoint discharges (like runoff from urban areas or from farms), the origins of various toxic pollutants and their impacts on aquatic ecosystems, and the costs of various means of reducing discharges of different pollutants. As Dorcey[37] and other Westwater researchers argue, without such information, it is difficult to appraise the effectiveness of present institutions or assess the likely success of alternatives.

A second problem relates to the difficulties of attaining a coordinated approach to water pollution control in a situation where the federal and provincial governments both have responsibilities,[38] but where the focus for action is a region. The two governments jealously guard their prerogatives but both recognize that a coordinated effort is required. The Greater Vancouver Water District performs a useful role in domestic and industrial water supply, as well as in managing the sewerage and sewage disposal systems of that area but it does not have the technical capacity nor the political authority to handle a problem the scale of water pollution control in the lower Fraser.[39]

The third matter concerns the examination of alternative strategies for dealing with water pollution in the region. So far the emphasis has been on permits, fines, and subsidies. The permit and fine system seems to have been relatively ineffective to date.[40] Few offenders have been brought to court. Subsidies are available to both municipalities and industries to reduce the cost of pollution control, either through grants, loans, or income tax deductions. Unfortunately, these incentive schemes have not been integrated closely with requirements of pollution control agencies and so gaps have remained between what pollution control agencies desire and what subsidy programs have produced.[41] There are several other options that have received little attention. These include the imposition of fees on discharges that are permitted by the Pollution Control Branch, and sewer surcharges that would be levied on non-domestic dischargers who wished to dispose of wastes through such outlets. In both instances the charges could be so designated as to present the effluent producer with a choice: whether to pay the fee or find some other means of disposing of the waste.

Summary

The past century has witnessed some important shifts in the approach to water management in the Fraser River basin, particularly in the lower Fraser Valley. Progressively, it has moved from concentration on single purpose, locally focussed construction to multiple purpose schemes, and, more recently, to a search for multiple means.[42] Stress seems to have played an important role in stimulating these changes in the case of both adjustment to floods and water pollution control. The nature of the stress, however, has differed. In the case of floods, there is clear evidence that the 1948 flood in the lower Fraser Valley and the 1972 flood in the Oak Hills area led to important changes in the approach to planning and policymaking. In the case of water pollution control, the environmental movement and the threat of federal government intervention seem to have been the dominant factors in stimulating institutional modifications.

Although there have been several major alterations in policies and administrative structures in the past five years, neither the flood problem nor that of water quality in the lower Fraser Valley has been satisfactorily solved. This is partly because the problems are not completely understood. It is also partly because no means have been found for reconciling the conflict between the maintenance of salmon runs and the construction of flood control dams. There are also uncertainties as to whether such alternatives as incentive schemes would be more effective than existing strategies for dealing with water pollution. But perhaps most importantly, the public does not presently perceive the need for action with the same urgency as the scientists, the experts, or the environmentalists. Until they do, reliance will continue to be placed on crisis as the signal and vehicle for action.

Notes

1. Henry Hart, "Crisis, Community, and Consent in Water Politics," *Law and Contemporary Problems,* Vol. 22, No. 3 (Summer 1957), pp. 510-537.

2. See, for instance, Helen Ingram, "Patterns of Politics in Water Resources Development," *Natural Resources Journal,* Vol. 11, No. 1 (January 1971), pp. 102-118; Robert J. Morgan, "Pressure Politics and Resources Administration," in James E. Anderson (ed.), *Politics and Economic Policy-making* (Read-

ing, Mass.: Addison-Wesley, 1970), pp. 390-413; and Matthew Holden Jr., "Political Bargaining and Pollution Control," in Anderson (ed.), *Politics and Economic Policy-making,* pp. 435-458.

3. David K. Todd, *The Water Encyclopedia* (Port Washington, N.Y.: Water Information Center, 1970), p. 119.

4. Fraser River Board, *Final Report on Flood Control and Hydroelectric Power in the Fraser River Basin* (Victoria, B.C.: Queen's Printer, 1963), p. 18.

5. B.C. Department of Economic Development, *The Manual of Resources and Development* (Victoria, B.C., November 1974), p. 23.

6. B.C. Department of Economic Development, *The Manual of Resources and Development,* p. 25.

7. M. A. Micklewright, "Problems in Canadian West Coast Fisheries: Government Policy and Intervention," in M. C. R. Edgell and B. H. Farrell (eds.), *Themes on Pacific Lands,* Western Geographical Series, Vol. 10, University of Victoria (1974), p. 260.

8. There are ninety-one Indian bands with a total population of 18,000 on lands in the Fraser River basin. Phillip A. Meyer, *Recreational and Preservation Values Associated with the Salmon of the Fraser River* (Vancouver, B.C.: Fisheries and Marine Service, Environment Canada, 1975), p. 5.

9. Meyer, *Recreational and Preservation Values;* see also P. Pearse and G. Bowden, *The Value of Fresh Water Sport Fishing in British Columbia* (Victoria: B. C. Fish and Wildlife Branch, 1971).

10. Fraser River Board, *Preliminary Report on Flood ·Control and Hydro-Electric Power in the Fraser River Basin* (Victoria, B.C.: Queen's Printer, 1958), p. 123.

11. B.C. Department of Lands and Forests and Water Resources, *Power in British Columbia* (Victoria, B.C., 1974), p. 14.

12. K. E. Patrick, "The Water Resources of British Columbia," in *Inventory of the Natural Resources of British Columbia,* (Victoria: B.C. Natural Resources Conference, 1964), p. 96.

13. Fraser River Board, *Preliminary Report,* p. 153.

14. Pearse and Bowden, *Value of Fresh Water Sport Fishing.*

15. Bruce Hutchison, *The Fraser* (Toronto: Clarke Irwin and Co., 1950), pp. 3-5.

16. W. R. Derrick Sewell, *Water Management and Floods in the Fraser River Basin,* University of Chicago Department of Geography Research Series No. 100 (Chicago, 1965), p. 21.

17. Discharges at Mission are generally expressed in terms of stage, since effects of tides are felt at that point. These effects are not present at Hope. Rough calculations are made, however, of actual flows at Mission by extrapolation from Hope readings.

18. Fraser River Board, *Preliminary Report,* p. 47.

19. Fraser River Board, *Final Report,* p. 21.

20. Hutchison, *The Fraser,* p. 347.

21. Fraser River Board, *Preliminary Report,* p. 50.

22. Fraser River Joint Advisory Committee, *Fraser River Upstream Storage Review Report* (Victoria, B.C., 1976), p. 9.

23. Fraser River Board, *Final Report.*

24. Fraser River Joint Advisory Committee, *Fraser River Upstream Storage Review Report.*

25. Ibid., p. 35.

26. Ibid., p. 9.

27. Anthony H. J. Dorcey, "Implementing Pollution Control," in Anthony H. J. Dorcey (ed.), *The Uncertain Future of the Lower Fraser* (Vancouver: University of British Columbia Press, 1976), p. 125.

28. Kenneth J. Hall, "The Quality of Water, The Lower Fraser and Sources of Pollution," in Anthony H. J. Dorcey (ed.), *Uncertain Future,* p. 38.

29. Dorcey (ed.), *Uncertain Future.*

30. Canadian Council of Resource Ministers, *Background Papers and Proceedings of the National Conference on Pollution and Our Environment,* Montreal, Quebec, November 1966.

31. David Estrin, "Tokenism and Environmental Protection," in O. P. Dwivedi, *Protecting the Environment: Issues and Choices—Canadian Perspectives* (Toronto: Copp Clark Publishing Co., 1974), pp. 123-138.

32. Denis Bellinger, "Canadian Water Management Policy Instruments: A National Overview," in Bruce Mitchell (ed.), *Institutional Arrangements for Water Management: Canadian Experiences,* University of Waterloo Department of Geography Publication Series No. 5 (Waterloo, Ontario, 1975), pp. 1-42; Steven Schatzow, "The Influence of the Public on Federal Environmental Decision-Making in Canada," in W. R. Derrick Sewell and J. T. Coppock (eds.), *Public Participation in Planning* (London: John Wiley and Sons, December 1976); and O. P. Dwivedi, *Protecting the Environment.*

33. Hall, "The Quality of Water," p. 38.

34. L. M. Hoos and G. A. Packman, *The Fraser River Estuary, Status of Environmental Knowledge to 1974,* Special Estuary Study Series No. 1, West Vancouver, B.C.: Environment Canada, 1974).

35. Mary L. Barker, *Water Resources and Related Land Uses: Strait of Georgia–Puget Sound Basin,* Geographical Paper No. 56 (Ottawa: Environmental Canada, Lands Directorate, 1974).

36. Hall, "The Quality of Water," p. 45.

37. Dorcey, *Uncertain Future,* p. 146.

38. The federal government, under the British North America Act, 1868, has responsibilities for the protection of fisheries, navigation, and harbor development, as well as an overall obligation to protect the national interest. The provincial government, under the same act, owns the resources within its territory and has responsibility for the health and welfare of its citizens.

39. Mark H. Sproule-Jones and Kenneth G. Peterson, "Pollution Control in the Lower Fraser: Who's in Charge?," in Dorcey (ed.), *Uncertain Future,* p. 171.

40. Dorcey, *Uncertain Future,* p. 147.

41. Ibid.

42. For a detailed discussion of the evolution of these concepts, see Gilbert F. White, *Strategies of American Water Management* (Ann Arbor: University of Michigan Press, 1969).

7

The Plate River Basin

J. C. Day

An integrated river basin management program was initiated in the mid-1960s in the Plate River basin by Argentina, Bolivia, Brazil, Paraguay, and Uruguay. Cooperative international actions were intended to achieve balanced and harmonious basin development and accelerate regional economic advances.

This chapter reviews the basin development efforts. It begins with a survey of basin characteristics, and then considers major thrusts in the cooperative development program to date, concluding with an examination of the degree to which program impacts have been traced by the engineers and scientists engaged in planning for the Plate basin. Emphasis is placed on studies which are worthy of additional research to determine the effects of river and resource development.

Basin Characteristics

The Plate basin comprises four sub-basins—the Parana (49 percent), the Paraguay (35 percent), the Uruguay (12 percent), and the Plate (4 percent)—which cumulatively drain more than

J. C. Day is associate professor, Department of Geography, University of Waterloo, Waterloo, Ontario. Citizens of Argentina, Bolivia, Brazil, Paraguay, and Uruguay, and employees of international agencies supplied the information on which this chapter is based. Newton V. Cordeiro, Organization of American States (OAS) was especially helpful in discussing many issues considered here. J. G. Nelson reviewed parts of the manuscript and made valuable suggestions for its improvement.

FIG. 7–1 THE RIVER PLATE BASIN

3 million km[2]: an area equivalent to 17 percent of Latin America or 40 percent of the coterminous United States[1] (Figures 7-1 and 7-2). Five nations share the basin: 44 percent of its area is in Brazil, 32 percent in Argentina, 13 percent in

Paraguay, 6 percent in Bolivia, and 5 percent in Uruguay. Expressed in terms of the proportion of each nation within the Plate drainage, all of Paraguay, 80 percent of Uruguay, 37 percent of Argentina, 19 percent of Bolivia, and 17 percent of Brazil are involved.

The Rivers

The Parana is a major river with a total length of 4,262 km and with its longest reach (2,965 km) in the upper basin above the Paraguay River confluence. The Paraguay extends southward 2,935 km from its union with the Amazon system through southern Brazil, Paraguay, and northern Argentina to the point where it joins the Parana. Comparatively, the Uruguay River is much smaller than the other tributaries. In its upper 788 km reach it traverses Brazilian territory; in the next 815 km section it forms the boundary between Argentina and Brazil; and its lowest 463 km constitutes the Uruguay-Argentina frontier.[2] The Plate River extends eastward from the Parana delta to a line joining Punta del Este, Uruguay, and Cape San Antonio, Argentina.

Discharge. The Parana River dominates discharge from the Plate basin. Fed primarily by heavy precipitation (in places exceeding 2,400 mm/year) on the western flanks of the Brazilian coastal mountains which form the eastern basin boundary (Figure 7-3), and other large areas which receive in excess of 1,400 mm of rainfall annually (Figure-7-4), mean annual runoff is comparatively heavy (Figure 7-5) and discharge at Posados is 11,800 m^3/sec (Figure 7-6). Precipitation is lighter and evapotranspiration higher in the Paraguay basin than the Parana, producing considerably lower mean annual runoff. In the Pilcomayo headwaters of the Bolivian Andes, precipitation exceeds 1,000 mm annually. A large north-south area paralleling the Andes Mountains, and the Paraguay–lower Parana rivers, receives less than 1,000 mm annually. Slightly heavier rainfall (1,000–1,600 mm/yr) is experienced in the northeastern part of the Paraguay basin. As a result the average annual

126

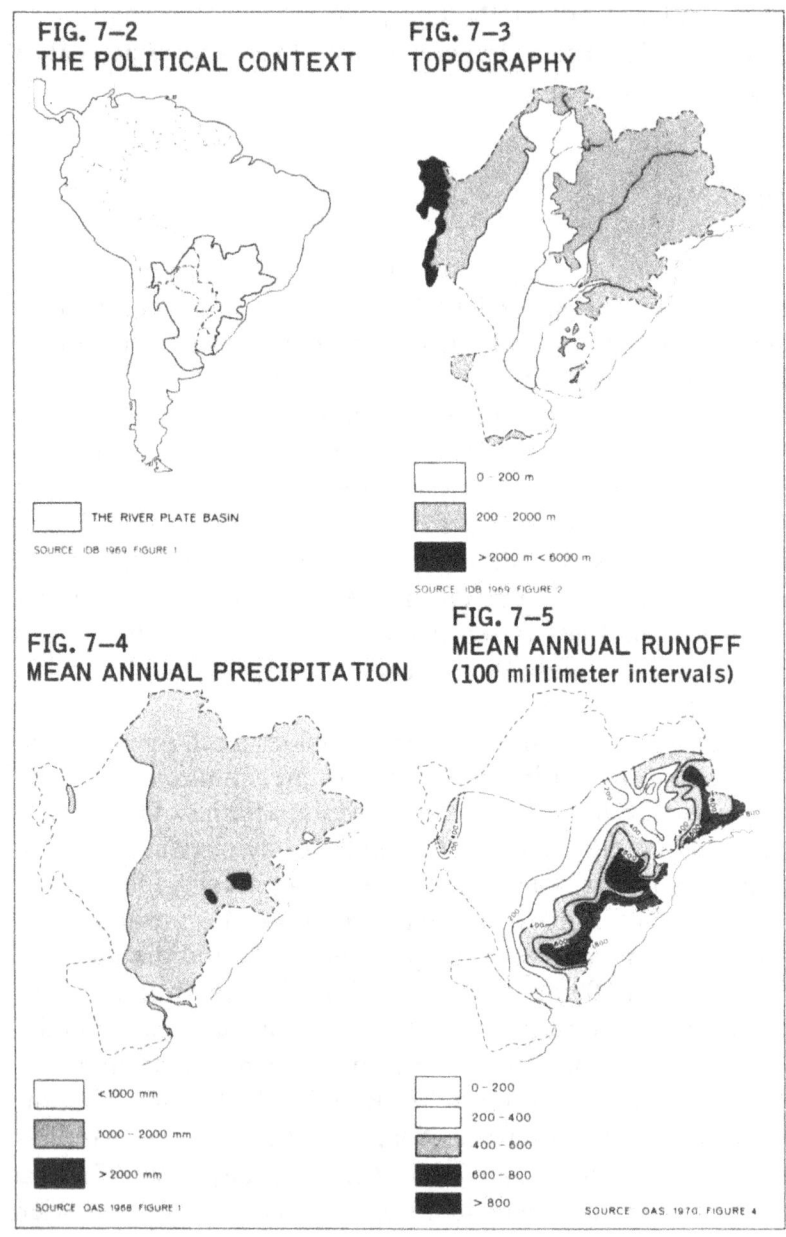

FIG. 7–2
THE POLITICAL CONTEXT

THE RIVER PLATE BASIN

SOURCE IDB 1969 FIGURE 1

FIG. 7–3
TOPOGRAPHY

0 - 200 m

200 - 2000 m

>2000 m < 6000 m

SOURCE IDB 1969 FIGURE 2

FIG. 7–4
MEAN ANNUAL PRECIPITATION

<1000 mm

1000 - 2000 mm

>2000 mm

SOURCE OAS 1968 FIGURE 1

FIG. 7–5
MEAN ANNUAL RUNOFF
(100 millimeter intervals)

0 - 200

200 - 400

400 - 600

600 - 800

> 800

SOURCE OAS 1970 FIGURE 4

flow of the Paraguay at Asuncion is 2,940 m³/sec, the Paraguay at Corrientes is 15,860 m³/sec, and at Rosario the main channel flow is 14,457 m³/sec. In comparison, the Uruguay River discharge is about 5,400 m³/sec at Salto, Uruguay, fed mainly by heavy precipitation in its Brazilian headwaters.[3]

The magnitude of peak discharges and areas of flooding throughout the basin under unregulated conditions is imperfectly understood. In its extensive review of physical conditions, the Organization of American States (OAS) was unable to compile sufficient historical data to prepare a flood hazard map. Indeed, in many instances the historical maximum instantaneous flows are not available for stream gages of maximum record.[4] However, in two exceptional Parana River floods during 1904-1905 and 1965-1966, 4.5 million ha were reportedly inundated in Argentina and 0.6 million ha in Paraguay downstream from Posadas-Encarnacion.[5] Similarly, heavy annual flooding is experienced in the Brazilian Pantanal, a 50,000 km² Paraguay River floodplain, although the process, magnitude, and importance of the inundations are imperfectly understood.[6] Flooding is reported annually in many other subbasins.

Population

The large Plate basin population is growing rapidly. In 1970, Brazilians comprised 60 percent of the Plate population, Argentinians 29 percent, Uruguayans 5 percent, Paraguayans 4 percent, and Bolivians 2 percent (Table 7-1). Comparatively, all Paraguayans live in the Plate basin, 94 percent of Uruguayans, 77 percent of Argentinians, 39 percent of Brazilians, and 29 percent of Bolivians. Although the 1972 rate of population increase within the basin was less than the 2.8 percent Latin American 1965–1970 average (Argentina 1.5 percent, Bolivia 2.6 percent, Brazil 2.9 percent, Paraguay 2.4 percent, Uruguay 1.3 percent),[7] nevertheless, the 1970 basin population of 61 million is expected to reach 80 million by 1980.[8] If the area contiguous to the basin is considered, particularly the

128

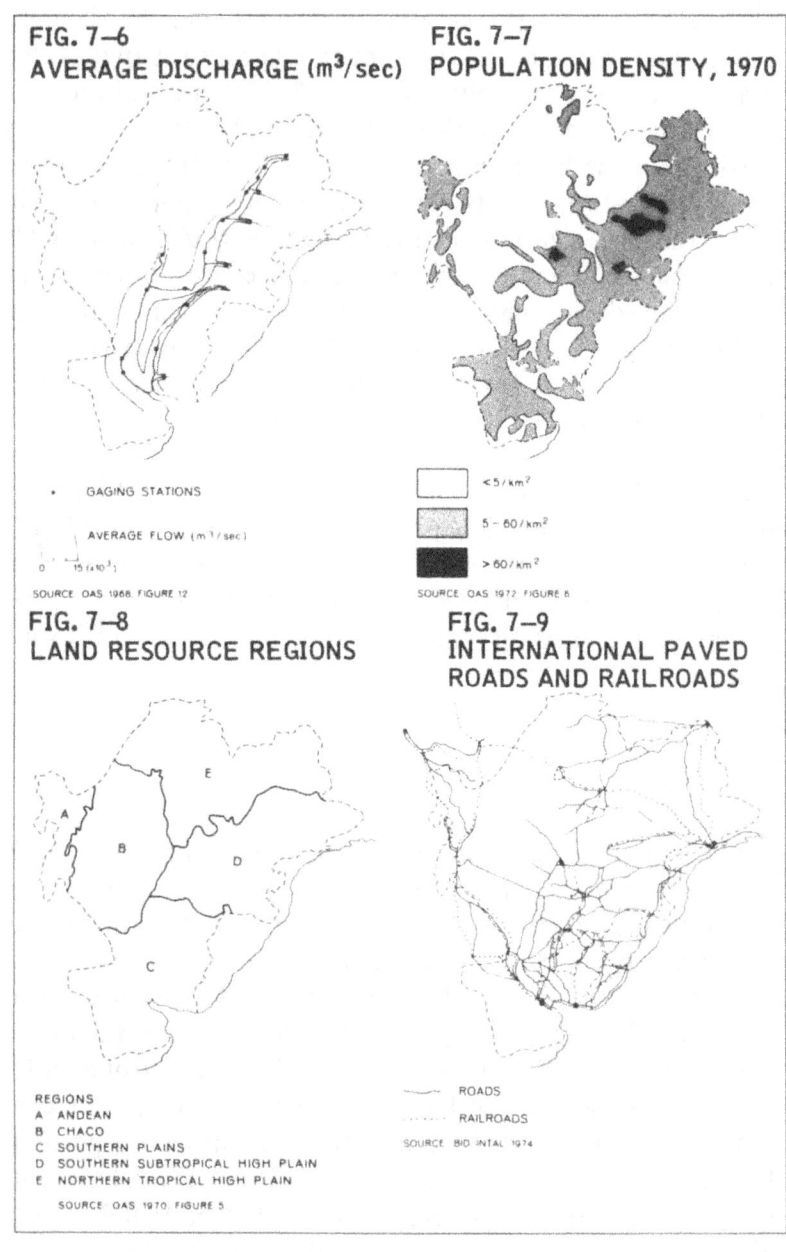

FIG. 7—6
AVERAGE DISCHARGE (m³/sec)

• GAGING STATIONS

AVERAGE FLOW (m³/sec)

0   15 (x10³)

SOURCE OAS 1968 FIGURE 12

FIG. 7—7
POPULATION DENSITY, 1970

< 5 / km²

5 – 60 / km²

> 60 / km²

SOURCE OAS 1972 FIGURE 6

FIG. 7—8
LAND RESOURCE REGIONS

E

A

B

D

C

REGIONS
A   ANDEAN
B   CHACO
C   SOUTHERN PLAINS
D   SOUTHERN SUBTROPICAL HIGH PLAIN
E   NORTHERN TROPICAL HIGH PLAIN

SOURCE OAS 1970 FIGURE 5

FIG. 7—9
INTERNATIONAL PAVED
ROADS AND RAILROADS

——— ROADS
········· RAILROADS

SOURCE BID INTAL 1974

TABLE 7-1

PLATE BASIN POPULATION CHARACTERISTICS: 1970[a]

Country	1970 Population (thousands)			Population in Cities of 100,000 or more[b]			Basin Population Density/km[2] 1970 (rural & urban)
	National	Basin	% in Basin	National	Basin	% in Basin	
Argentina	23,451	18,016	77	12,737	10,615	78	18.1
Bolivia	4,914	1,434	29	622	103	6	7.1
Brazil	93,048	36,664	39	22,947	8,603	37	25.9
Paraguay	2,406	2,406	100	374	374	100	6.0
Uruguay	2,886	2,706	94	1,783	1,709	96	19.7
Totals	126,705	61,226	48	38,463	20,904	54	19.4

[a]From: Organización de los Estados Americanos, 1969d, 333.72-S-7901, Mapa 6.

[b]Based on most recent census: Argentina, 1960; Bolivia, 1950; Brazil, 1960; Paraguay, 1962; Uruguay, 1963.

eastern Brazilian seaboard (including Rio de Janeiro which has strong economic linkages with the Plate area) the population is much larger.

A megalopolis is taking shape within the southern cone of Latin America continuous with and including the basin. Brasilia (512,000) and nearby Anapolis (96,000) and Goiania (222,000) form an isolated population concentration in the northeast. Rio de Janeiro (4,131,000) and Sao Paulo (9,000,000) are growing rapidly with heavy settlement arteries extending westward to the Parana River. South of Sao Paulo the cities of Curitiba (889,000), Ponta Grossa (146,000), and others east of the basin are also expanding. Contiguous to the Plate River, Montevideo (1,346,000), Buenos Aires (8,309,000), and La Plata (382,000) form the southern end of the developing eastern population axis. Another primary settlement artery extends up the Parana River through Rosario (689,000), Parana (141,000), Sante Fe (243,000), Corrientes (151,000), Resistencia (156,000), Posadas (140,000), and up the Paraguay River to Asuncion (397,000). A secondary line of settlement follows the Uruguay River. In the western basin the population is concentrated along transport lines linking Salta, Argentina (179,000), and Sucre, Bolivia (48,000). However, there are still massive areas, particularly in the north and central basin, where population density is less than 5 people/km² although the basin average is 19.4/km² (Figure 7-7). Roughly half of the basin population resides in cities greater than 100,000 inhabitants (Table 7-1).

Industries and Agriculture

The level of industrialization is highly variable among the basin nations. Argentina and Brazil have the only large industrial complexes which accounted for roughly 35 percent of the gross domestic product of both nations during 1971. The comparative amount is 23 percent in Uruguay, 20 percent in Paraguay, and 13 percent in Bolivia. Traditionally, Argentinian industries processing agricultural products have been most important. Government efforts to stimulate production of nondurable consumer goods were not effective. But durable goods industries have had some success in reaching a competitive world position. Brazil experienced rapid growth in the metallurgical, chemical, transport equipment, and capital goods industries during the 1960s and early 1970s. Industries in the three smaller nations mainly concentrate on some aspects of agriculture or mining; however, petroleum finds in Bolivia created a fast-growing refining industry in the 1970s.

Agriculture is fundamental to the economies of all basin nations. Nevertheless, in recent years only Brazil enjoyed progressive expansion of this sector enabling its agriculture to supply low-cost food, industrial inputs, and exportable surpluses. Although Argentinian agriculture furnishes 74 percent of all exports, growth of this sector has been slow. While Paraguayan agriculture furnishes 32 percent of the gross domestic product, employs 60 percent of the labor force, and provides 95 percent of exports, it is also growing slowly. And in Uruguay this important economic sector declined in importance in the seventies, contributing to that nation's rapid inflation and economic stagnation.[9]

Land Resource Regions

The OAS subdivides the basin into five land resource areas (Figure 7-8). With the exception of its lowest part, the *Andean region* has a cold and dry climate, rough topography, extensive rocky areas, and poor soils. Most agriculture is subsistence and agricultural land use intensity is low.

Development potential in the Argentinian, Paraguayan, and Bolivian *chaco* is generally restricted due to aridity, poor soils, flood hazards, and extreme temperature fluctuations. Although cotton, corn, sugar cane, fruits, and other crops are raised locally, extensive cattle ranching is most common, with some forestry in the extreme eastern area.

The most intensive wheat and cattle-raising area in the basin and in South America is located on the *southern plain.* Its humid temperate climate, deep, well-drained soils, and flat topography are suitable for the production of a variety of grains, vegetables, cereals, oils, and fibers. Natural limiting factors include local inadequate drainage, flooding in the Parana delta and floodplain, and periodic drought in the southeastern area.

The high northern plains are differentiated into two zones principally on the basis of climate, vegetation, and geology. The *southern subtropical region* is comprised mainly of rough, forested mountains where wood, pulp, and paper are important. Brazilian coffee production is concentrated in favorable microclimatic zones of this area and rice, tobacco, cotton, yerba mate, sugar cane, fruits, and cereals are also locally grown. Severe erosion has been experienced in urban and agricultural areas. Agricultural land use intensity in the *northern tropical region* is low as a result of environmental constraints, distance from large population centers, and inadequate infrastructure. At present, agriculture is restricted to extensive cattle raising and limited production of corn, coffee, tobacco, vegetables, and sugar cane.[10]

Transportation and Trade

In the late 1960s, trade among the basin countries was dominated by Argentinian-Brazilian ocean movements. On a smaller scale, river transport between Argentina and Uruguay, and Argentina and Paraguay, was also important. Nevertheless, exchange among the five nations was generally modest. This situation basically stems from the fact that national transportation systems in, and peripheral to, the Plate basin evolved as radial systems which link production areas with coastal or river transshipment points in each nation to world markets. Although interconnections between basin nations traditionally had low priority, recent innovations are changing this situation and trade between basin nations could expand dramatically in the future.[11]

Transportation is the river use of greatest historical importance. Initially the Plate system served as the most easily accessible route northward into the basin from the Atlantic. But in recent decades, its rivers served mainly to transport heavy and bulky cargoes of low unit value.

The 4,200 km primary waterway network includes: the Paraguay from its mouth to Corumba, Brazil; the Parana from its mouth to Port Mendes at Guaira Falls, Brazil; Uruguay from its mouth to Salto, Uruguay; the Plate from the Atlantic along dredged channels to Montevideo, to Buenos Aires, and to its principal tributaries. Navigation in this system is limited to shallow-draft ocean ships in the lower reaches of the system and shallower-draft riverboats and barges in the upstream areas. Navigation is difficult because of variable water depths, numerous shallow reaches, and the shortage of modern navigational aids throughout the system. In addition, 4,900 km of less dependable, lower-capacity secondary river routes exist on tributaries or further upstream.[12]

In the lower Parana, oil, sand, and grain are the main commodities. More than 90 percent of Argentinian imports and 75 percent of their exports are carried on the Plate River.[13] An alternative deep-water transshipment port to service the Plate basin is under study in Argentina to replace Buenos Aires because of siltation problems. Further upstream, 85 percent of Paraguay's foreign commerce is handled at Port Asuncion on the Paraguay River. It is possible that navigation locks associated with hydroelectric developments throughout the Plate system, as well as port, channel navigation, and river fleet improvements, will strengthen the role which the river plays in increased foreign commerce by Plate River nations in the future.[14]

The few good road connections between Plate basin countries are recent, and many links are under construction (Figure 7-9). Recent international highways revolutionized trading patterns among the five basin nations. A road connection between Brazil and Paraguay permitted 84 percent of Paraguayan exports by value to be carried to Brazil in trucks in 1971. Similarly, trucks carried 55 percent of the exports from Uruguay to Brazil in 1970. Conversely, the virtual absence of trade between Bolivia and Paraguay is due to the fact that these nations are linked best by air. The transchaco route connecting Asuncion to Bolivia has not yet been paved.[15] A highway network to connect Bolivia with its five neighbors is being considered. The fact that only 7–15 percent of the most important bilateral trade between Argentina and Brazil is by road reflects the poor highway connections between these nations.[16]

Although railroads were the first interconnections among basin nations, the importance of international trade carried by trains is low and declining in contrast to other forms of transport. In some instances, as between Bolivia and Brazil, and Brazil and Paraguay, connections were never made (Figure 7-9). In others, incompatibility of track widths between Argentina and Brazil, and Brazil and Uruguay, necessitates border cargo transshipments. There is a possibility that new bridges between the basin riparians will increase railroad traffic in the future.[17] A route linking Antofagasta, Chile, northwestern Argentina, Bolivia, and Corumba, Brazil, was used experimentally in 1974.[18]

The Development Program

Cooperative actions in the Plate program are less than nine years old. Understandably there are few visible effects directly attributable to basin programs. An early and continuing interest relates to the fact that little detailed knowledge was available concerning physical, human, and infrastructure resources of the Plate. The Department of Regional Development, OAS, synthesized and analyzed existing hydrological and climatological data and recommended a program for expansion of the basin hydrometerological network.[19] They also inventoried existing aerial photographic and map resources and compiled preliminary analyses and maps of soils, geology, vegetation, groundwater, agriculture, stock raising, population, energy, and transportation conditions throughout the basin.[20] The Institute for Latin American Integration of the Inter-American Development Bank (IDB) prepared similar analysis of basin resources.[21]

More recently, the OAS undertook or completed regional development and pre-feasibility studies of specific areas in cooperation with basin countries and the United Nations Development Programme (UNDP). These include: the Santa Lucia basin, Uruguay; the northeastern Parana State, Brazil; the upper Bermejo basin of Argentina and Bolivia; the lower Bermejo basin of Argentina, Bolivia, and Paraguay; the Aquidaban project, Paraguay; the Pilcomayo River basin of Argentina, Bolivia, and Paraguay; and the upper Paraguay pantanal region of Brazil (Figure 7-10).[22]

United Nations agencies have been active in assisting basin nations. The UNDP has worked since June 1969 to enhance Parana River navigation with interest in improvements to channels, ports, and river fleets; socioeconomic hinterland studies; and hydraulic, ecological, and water quality conditions.[23] Further upstream, the UNDP assisted with navigation improvements between Asuncion and the Paraguay-Parana confluence.[24] Studies in reaches of the lower Parana and Paraguay rivers were financed in part by the IDB.

Considerable attention is being directed to the Pantanal, a large floodplain with a drainage area of 400,000 km^2 in the

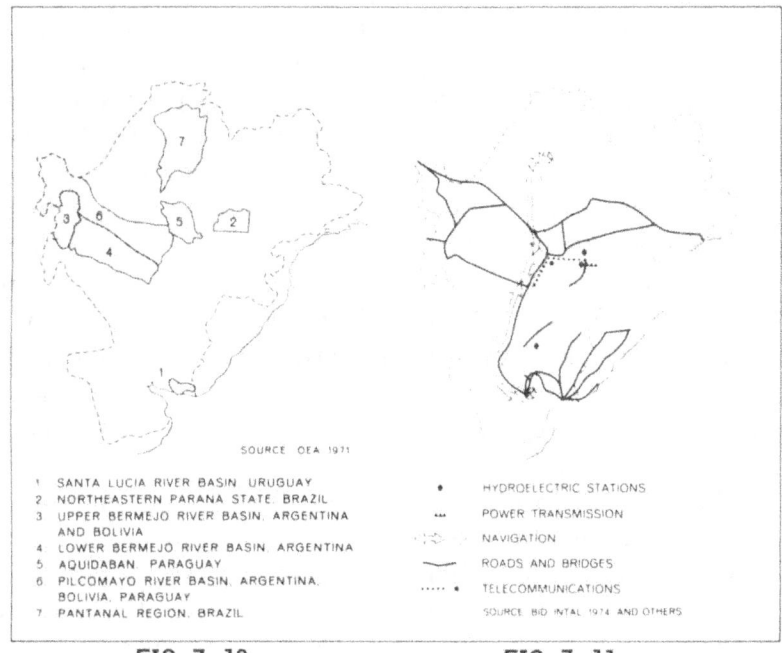

SOURCE OEA 1971

1 SANTA LUCIA RIVER BASIN. URUGUAY
2 NORTHEASTERN PARANA STATE. BRAZIL
3 UPPER BERMEJO RIVER BASIN, ARGENTINA
   AND BOLIVIA
4 LOWER BERMEJO RIVER BASIN, ARGENTINA
5 AQUIDABAN. PARAGUAY
6 PILCOMAYO RIVER BASIN, ARGENTINA.
   BOLIVIA. PARAGUAY
7 PANTANAL REGION. BRAZIL

●   HYDROELECTRIC STATIONS
...   POWER TRANSMISSION
      NAVIGATION
⌐   ROADS AND BRIDGES
..... ● TELECOMMUNICATIONS
      SOURCE BID INTAL 1974 AND OTHERS

FIG. 7–10
REGIONAL
DEVELOPMENT STUDIES

FIG. 7–11
PHYSICAL
INTEGRATION PROJECTS

Paraguay River basin, of which 350,000 km² are in Brazil
(Figure 7-10). The United Nations Economic and Social
Council conducted a survey to provide basic knowledge of
hydraulic resources of the area from 1967-1972 including
rainfall, overland flow patterns, flooding and flood forecasting, and a mathematical model of the upper Paraguay basin.[25]
In 1974, the Brazilian government allocated $650 million to its
new Special Pantanal Development Fund for the 1974-1976
period to improve transportation, regulate streamflow to
control flooding, improve navigation and sanitation, expand
energy resources, improve agriculture, and promote industrialization. There is a possibility that Brazil will be assisted in
this work by the UNDP and the OAS.

The United Nations supported systematic study and exploitation of the Plate basin fishery resources. Sigler suggested
ecological and other studies to assist in the rational exploita-

tion of fishery stocks.[26] Subsequently, D'Alarcao and Lagler
conducted a preliminary investigation of basin fishery resources which they considered broadly in terms of three ecosystems: Atlantic, Plate River, and inland rivers. They found that
the substantial and diverse fishery is poorly understood and
largely undeveloped. Concurrently, seventeen development
projects valued at approximately $2.5 billion were underway
with fisheries implications. Such works, affected by industrial
activities pollution, and siltation, were deteriorating habitats.
Diversion projects, in particular, were identified as being
particularly dangerous due to the potential invasion of undesirable plants, fish, and other organisms. They recommended a
broad range of institutional arrangements, international agreements, basic studies, and personnel to permit rational development of the basin fishery resources.[27]

Human health problems have been discussed internationally
by basin nations for decades. In four meetings of the health
ministers from 1948 to 1961, measures to prevent a variety of
frontier epidemics were considered.[28] Subsequently, the Pan-
American Health Organization (PAHO) conducted a preliminary reconnaissance of surface- and groundwater quality conditions, areas of natural, urban, and industrial and synthetic
organic pollutants, and the interrelationships of water quality,
health, and hydraulic works. PAHO recommended immediate
sequel studies to complete its preliminary basin water quality
inventory and a plan to control water quality throughout the
basin.[29] But as yet, further cooperative actions have not been
undertaken.

International Development Assistance

It is difficult to determine the level of development assistance
provided by international organizations and banks. Usually
loans to basin countries are related to sectors of the economy
rather than the Plate program and there is no estimate of the
degree to which transportation improvements, electricity
transmission, and telecommunications are located within or
outside the basin. And few international agencies keep sum-

mary accounts of basin regional development expenditures. Moreover, many of these improvements have not been approved collectively by the foreign affairs ministers of the Plate basin nations.

International assistance to the mid-1970s focusses on electricity generation and transmission, and transportation improvements (Figure 7-11). These items compromise nearly 90 percent of the $12 billion assistance made available by the IDB, the World Bank group, and the United Nations (Table 7-2). The assistance has been received primarily by Paraguay, Argentina, and Brazil. These data do not reflect substantial expenditures in the basin by the OAS, the U.S. Agency for International Development, and many other agencies which do not report their actions in the format used in this report. Available information does not permit an estimate of the order of magnitude of such expenditures.

Program Impacts

For a variety of reasons little is known about Plate basin development impacts. Most important is the short program life. Less than a decade has elapsed since the five Plate countries agreed to cooperate in the development of their shared resources and few projects to date have been sanctioned, or undertaken cooperatively, by the basin states.

Most of what is known concerning the effects of earlier national and binational management actions concerns alleged adverse influences on the downstream riparian, Argentina. Particularly in provincial Argentinian newspapers, attention is drawn to a variety of circumstances which could negatively affect their national welfare. Existing and potential diversions which take water from the Parana and Uruguay rivers would reduce hydroelectric potential downstream.[30] A similar effect would be induced by proposed Brazilian irrigation projects which decrease discharge through increased consumptive water use.[31] Both of these plans could negatively affect shipping interests by lowering or raising water levels in navigation channels over the long and short term. Moreover, it is alleged

TABLE 7-2

PLATE BASIN PHYSICAL INTEGRATION PROJECTS: THE INTER-AMERICAN
DEVELOPMENT BANK,[a] THE UNITED NATIONS,[a] AND THE
WORLD BANK GROUP[b]: 1950-1975
($US x 10³)

Development Process	Argentina	Bolivia	Brazil	Paraguay	Uruguay	Total
Water Transport	54,715	4,148	264,300	1,680	1,272	326,115
Electricity, water regulation, and transmission	2,991,874	47,184	768,983	4,545,068	543,380	9,222,604
Transportation	661,500	87,205	1,325,150	167,314	72,978	2,314,147
Telecommunications	480	--	--	15,400	--	15,880
Industry	6,600	400	86,000	--	35,000	128,000
Education	--	--	--	5,100	--	5,100
Minerals	11,000	6,200	22,000	--	--	39,200
Agriculture	15,000	7,500	53,300	58,600	71,500	245,100
Planning	--	--	--	4,000	--	4,000
Water supply, sewage	--	--	73,000	--	--	73,000
Totals	3,741,169	152,637	2,592,739	4,797,162	724,130	12,007,837

[a]Expenditures are estimated to 1973 from: Banco Interamericano de Desarrollo, 1972; Argentina, Ministerio de Relaciones Exteriores y Culto, y Banco Interamericano de Desarrollo..., 1973, Vol. II.

[b]Compiled from news releases of the World Bank Group--the International Development Association, the International Bank for Reconstruction and Development, and the International Finance Corporation--1949-75.

that at Posadas, Argentina, it is increasingly difficult to find coarse river sand for the local construction industry since reservoir construction began in Brazil.[32] Ecologists indicate that river environment changes produced by upstream dams could negatively affect fish populations downstream.[33] Medical doctors fear that the settlement process which accompanies new dam construction could lead to the diffusion of water-borne diseases into Argentina from Brazil if adequate sewage facilities are not constructed.[34]

Finally, there is concern about potential transboundary

pollution from Brazil as development occurs throughout the Plate basin and possible flood damage to downstream communities if Brazilian dams should fail.[35] Although it has been impossible to verify any of these allegations, the diversity of concerns indicates the degree of uncertainty which the downstream riparians in Argentina feel about upstream developments, particularly those in Brazil.

Many of the existing large projects were financed by the IDB and the World Bank group and to a lesser degree by a variety of other interests. Although they have not been adopted formally as part of the basin program, ultimately they will contribute in a major way to the infrastructure, available resources, trade, prosperity, and the degree of international harmony shared by basin countries. Some aspects of international assistance relating to hydroelectricity generation, transportation, and agricultural productivity span more than two decades of experience. But little is known about the effects of such developments other than the primary purposes for which the actions were taken: kilowatt hours produced, traffic flows, hectares cultivated, and crop yields. In most instances, the induced biophysical and social effects of these actions have not been documented.

Factors which contribute to this situation are numerous. In general, as throughout the rest of the world, there are inadequate biophysical bench mark data in the basin from which project-induced perturbations may be assessed. Inventories of physical resources such as geology, soils, water quality, plants, animals, and flood and erosion hazards are not sufficiently precise to permit measurement of development changes. Just as important, however, is the fact that the process of monitoring project impacts has not been institutionalized. International financial institutions do not require that biophysical and social data be collected prior to development nor that effects be monitored after project completion as a condition for obtaining loans. As a result, little precise information on the full range of development effects is available and a great deal of potentially valuable experience to guide future work has been lost.

In spite of these unknown aspects of development, the OAS

reports have documented conditions and hazards as precisely as time and funds permit in their basin inventory maps and seven regional development studies. Projects of the UNDP concerning Parana and Plate river navigation, the FAO fisheries assessments, and other studies include social and biophysical surveys which vary widely in the level of detail at the time of the agency involvement, but these assessments are not ongoing. Basin countries conduct similar studies infrequently, as time and funding allow. For example, Electrobras currently is monitoring physical parameters in the vicinity of some of its hydroelectric projects.

More recently, the United Nations Environment Programme (UNEP) proposed a world-wide environmental assessment to promote integrated, ecologically sensitive development efforts. This would entail comprehensive appraisal of problems associated with human settlements, human and environmental health, terrestrial ecosystem management, development effects, energy production, and natural disasters.[36] Efforts are being made to incorporate UNEP policies for environmental monitoring and ecosystem preservation in the Paraguay River Pantanal program and possibly other basin areas.

In summary, while little precise data concerning Plate basin development impacts are currently available, there is an enormous opportunity to begin to trace development effects when future cooperative programs are implemented. Biophysical parameters such as soils, stream channels, river discharge, surface- and groundwater quality, and flora and fauna will be disturbed, and existing social and economic conditions will be influenced in a variety of ways. The development of means and training of personnel to permit monitoring of such changes would facilitate detection and arresting of undesirable induced modifications early in the development process, and avoidance of undesirable effects in subsequent projects. The utility of institutionalizing biophysical and social monitoring of future national and international Plate development actions deserves careful consideration by the basin countries and international financial institutions.

Research Needs

The major River Plate basin development problems and research needs are similar to those recognized in other major world river systems. The problems and research deficiences are basically biophysical and social in nature, and a number of researchers have made recommendations for their amelioration or solution.[37]

A fundamental biophysical problem concerns the basin water balance. International agreement is necessary concerning water use priorities. And strategies must be found to arrange for the timing and amount of flows through the system to meet the needs of as many water users as possible. This is essential in light of the enormous capability that the completed and planned water-control structures will create to manipulate the basin hydrological regimen.

An adjunct problem pertains to water quality. Considerable work is required to specify realistic water quality standards and institutional arrangements for their enforcement. There is an associated need to define and protect representative terrestrial, marine, and freshwater ecosystems in the form of national parks and biological reserves, as well as cultural areas of national significance. Such areas would serve a valuable function for tourism, research, and training. They also provide permanent monitoring sites of representative samples of land and water ecosystems against which the degree of disturbance in modified environments may be judged.

From a social perspective, an underlying problem is how new biophysical information can be incorporated into improved management systems. How, for example, can better knowledge of the timing and volume of river flows be used effectively to promote decision making which will increase international harmony and social welfare? Any research contributing to better understanding of more effective means of allocating scarce resources among increasing numbers of users as the level of biophysical information increases could be highly significant. In this context, comparative studies of the effect of differing sets of local, regional, or national perceptions and attitudes concerning future basin development alternatives in a variety of institutional and economic settings would also be a potentially useful management tool.

Notes

1. Organization of American States, General Secretariat, *Program of Regional Development* (Washington, D.C.: OAS, 1974). This report summarizes OAS projects in the Plate basin.
2. Naciones Unidas, Instituto para la Formación Profesional e Investigaciones, *Rios y Canales Navegables Internacionales: Aspectos Financieros, Juridicos, e Institutionales de su Desarrollo* (Buenos Aires: UNITAE, 1971), pp. 160-162.
3. Organización de los Estados Americanos, Departamento de Asuntos Económicos, Unidad de Recursos Naturales, *Cuenca del Río de la Plata: Inventario de Datos Hidrologicos y Climatologicos,* 333.72-S-7739.2 (Washington, D.C.: Secretaria General de la OEA, 1969b), maps 1 and 12.
4. Organización de los Estados Americanos (1969b), *Cuenca del Río de la Plata*, map 14.
5. Ruben Naranjo et al. (eds.), *Paraná el pariente del mar* (Rosario, Arg.: Biblioteca Popular, 1973), p. 263.
6. L. Lisoni and E. Stretta, "Hydrological Studies of the Upper Paraguay River Basin (Pantanal Region, Mato Grosso State, Brazil)," *Nature and Resources,* Vol. 6, No. 1 (1973), pp. 10-13.
7. Inter-American Development Bank, *Economic and Social Progress in Latin America: Annual Report 1973* (Washington, D.C.: IDB, 1973), p. 86.
8. Banco Interamericano de Desarrollo, Programa BID–Cuenca del Plata, *Informe Preliminar Regional,* Vol. 1 (Buenos Aires: BID, 1969), p. 56.
9. Inter-American Development Bank, *Annual Report 1973*, pp. 112-305.
10. Organización de los Estados Americanos, Departamento de Asuntos Económicos, Unidad de Recursos Naturales, *Cuenca del Río de la Plata: Inventario Andisis de la Informa-*

ción Básica sobre Recursos Naturales, 333.72-S-7901 (Washington, D.C.: Secretaria de la OEA, 1969c), pp. 83-99.

11. Argentina, Ministerio de Relaciones Exteriores Culto, y Banco Interamericano de Desarrollo, Instituto para la Integración de America Latina, *Inventario de Proyectos de Integración Fisica en la Cuenca del Plata* (Buenos Aires: BID, 1973), p. 41.

12. Argentina, *Inventario de Proyectos de Integración Fisica en la Cuenca del Plata,* p. 25.

13. Argentina, Secretaria de Marina, Servico de Hidrografia Naval, *Levantamiento Integral del Area del Plata* (Buenos Aires, 1967).

14. Argentina, *Inventario de Proyectos de Integración Fisica en la Cuenca del Plata,* pp. 25-40.

15. A. Hecht, Personal correspondence (Waterloo, Ontario, June 1975).

16. Inter-American Development Bank, *Economic and Social Progress in Latin America: Annual Report 1972* (Washington, D.C.: IDB, 1973), pp. 24-27, 83.

17. Argentina, *Inventario de Proyectos de Integración Fisica en la Cuenca del Plata,* pp. 73-77.

18. T. R. Lee, Personal correspondence (Santiago, Chile: August 1975).

19. Organización de los Estados Americanos, Departamento de Asuntos Económicos, Unidad de Recursos Naturales, *Cuenca del Río de la Plata: Inventario de Datos Hidrologicos y Climatologicos,* 333.72-S-7739.1 (Washington, D.C.: Secretaria General de la OEA, 1969a); and Organización de los Estados Americanos (1969b), *Cuenca del Río de la Plata.*

20. Organización de los Estados Americanos (1969b and 1969c), *Cuenco del Río de la Plata.*

21. Banco Interamericano de Desarrollo, *Informe Preliminar Regional.*

22. Organization of American States, *Program of Regional Development.*

23. United Nations Development Programme, *Improvement of Navigation in the Parana River (Phase II),* DP/Projects/400 (Arg/73/023: November 1974).

24. Naciones Unidas, Programa para el Desarrollo, *Estudio de la Navegación del Río Paraguay al sur de Asunción: Argentina-Paraguay,* DP/UN/RLA-65-235/1 (New York: United Nations, 1974).

25. L. Lisoni and E. Stretta, "Hydrological Studies of the Upper Paraguay River Basin."

26. Naciones Unidas, Organización para la Agricultura y la Alimentación, Informe al Gobierno de la Republica Argentina sobre el Programa de Investigaciones y Desarrollo Pesgueros del Río Paraná," por William F. Sigler, Rep., FAO/UNDP/TA; 2628 (Rome: ONVA y A, 1969).

27. R. P. d'Alarcao and Karl F. Lagler, "Mission on Evaluation of Ichthyological Resources in the River Plate Basin," United Nations Development Programme, DS/SF/310/Reg. 187 (October-November 1969).

28. Naciones Unidas, Oficina Sanitaria Panamericana, Acuerdo Sanitario y Actas Finales, *Reuniones de Ministerios de Salud de la Cuenca de Plata* (1948).

29. Naciones Unidas, Organización Mundial de la Salud, Oficina Sanitaria Panamericana, *Calidad de Aguas en la Cuenca del Plata,* por Ing. Walter A. Castignino (Washington, D.C.: PAHO, 1969).

30. Isaac Franciso Rojas, *Interés Argentinos en la Cuenca del Plata* (Buenos Aires: Museo Social Argentino, 1969), p. 7.

31. Rojas, *Interés Argentinos,* p. 2.

32. "Las Represas Brasileñas sobre el Paraná y la Extracción de Arena en Misiones," *El Territorio* (Posadas, Arg., September 10, 1969).

33. Rojas, *Interés Argentinos,* p. 3.

34. Dr. Humberto R. Cabral, "La Preservación del Caudal y Potencia del Río Paraná," *La Nación* (Buenos Aires, April 25, 1973), pp. 8-10.

35. Rojas, *Interés Argentinos,* p. 25.

36. United Nations Environment Programme, *The Proposed Program: Note by the Executive Director,* UNEP/GG/31 (Nairobi, 1975).

37. United Nations, Department of Economics and Social Affairs, *Integrated River Basin Development,* E/3066/Rev. 1

(New York: UN, 1970); W. C. Ackerman, G. F. White, and E. B. Worthington (eds.), *Man-Made Lakes: Their Problems and Environmental Effects,* Geophysical Monograph 17, (Washington, D.C.: American Geophysical Union, 1973); National Academy of Sciences, Committee on Water, *Water and Choice in the Colorado Basin: An Example of Alternatives in Water Management* (Washington, D.C.: National Academy of Sciences, 1968); and G. F. White, *Strategies of American Water Management* (Ann Arbor: University of Michigan Press, 1969).

8

African Experience: Some Observations from Kenya

R. S. Odingo

In many developing countries, the major river basins have been the last to be developed because they were regarded as too difficult or costly for human settlement. In some tropical countries, rivers were the home of tropical diseases like malaria (transmitted by mosquitoes) and sleeping sickness (transmitted by tsetse flies). As population expanded in the more environmentally favored areas for human settlement, only the semiarid lands and the formerly neglected river basins remained to be developed.

It was not until the last few decades that river basins became a new focus for rapid economic development through capital-intensive projects involving swamp drainage, irrigation, and waterpower development. Most of these new projects relied on imported technology and were launched on the assumption that technological innovations can help to overcome most of the evident problems of the river basin environment. The Tana River basin in Kenya demonstrates these points and will serve to point out tendencies and problems associated with the development of river basins in other parts of tropical Africa.

The complex management of river basins in the more developed parts of the world is usually based on relatively long

R. S. Odingo is associate professor of geography, University of Nairobi, Nairobi, Kenya.

records of various aspects of the river and on carefully worked out and integrated plans aimed at obtaining maximum benefits from human intervention. This is not always the case with river basins in the less-developed parts of the world, where the necessary data base for such planning is either absent or thin and fragmentary; heavy reliance has had to be placed on incomplete data. Moreover, it is rare that well-coordinated plans covering whole river basins are made. Rather it is common to find piecemeal developments, some of which were carried out during the colonial period and others which have been executed in more recent years. Thus, from the very start, it is clear that the concept of integrated river basin development is more often an aim than a reality in many developing countries. Problems arise because of failure to pay attention to many aspects of river basin development which are often taken for granted in the more developed parts of the world.

As already implied, the approach to the development of river basins in developing countries has been in the form of large-scale capital-intensive projects financed by the International Bank for Reconstruction and Development (World Bank), or through bilateral or multilateral aid, aimed at solving immediate problems like food shortages (by the implementation of irrigation schemes) and energy needs (through construction of hydroelectric dams, etc.). Flood control, unless accompanied by an irrigation project, is usually regarded as unproductive and is rarely supported by such aid. But swamp drainage followed by an irrigation scheme, as in the case of the Yala Swamp in western Kenya,[1] is often regarded as productive and likely to attract financial support from donor agencies. Finally, what could be called the rationalization of the availability of water for use in a less fortunate part of a river basin (as in the case in Egypt and the Sudan, which led to the proposal to go ahead with the construction of the ecologically controversial Jonglei Canal) is only recently beginning to attract financial support from various sources.

It is unfortunate that many of the projects recently developed or now proposed for development have failed to adopt the integrated river basin development approach, since it is only through such an approach that harmonious development can be achieved in a given area.[2] In this paper an attempt will be made to point out the problems for river basin development programs patterned upon partial and inappropriate solutions in the developed countries that appear most frequently in the developing countries.

Data Collection Problems

Integrated river basin development requires a wealth of data—collected, in many instances, over long periods. Most developing countries lack sufficient and appropriate data necessary for integrated development. The problem is not only availability but reliability of data for planning purposes. Data on streamflow or flood frequency may be available, but often it is not given in a standardized format. Second, much of the data may be unreliable because of the methods by which it was obtained. For example, where data on evaporation, river discharge, and sediment load happen to be available, they may not be reliable enough for meaningful forecasting and conclusions.

The development of strong meteorological services in developing countries, which initially were meant largely to serve aviation, led to the accumulation of much good meteorological data. Also, there is now a hydrological network with well-kept records of shorter duration in some developing countries, including Kenya. The presence of acceptable topographic map coverage and aerial photo coverage of important river basins fills one major data need, for maps are an essential starting point for planning. The recent availability of earth-satellite photography may provide a less expensive and more comprehensive alternative to aerial photo coverage; the use of remote sensing techniques for analysis of the land potential in a whole river basin also should be closely looked into. The same technique may lead to the preparation of inexpensive soil maps (now lacking), which are a prerequisite for agricultural development. The soil surveys need to be accompanied by other information such as:

1. indices of soil erodibility;
2. salinity hazard, especially in semiarid environments;
3. erosion plot experiments;
4. soil moisture available for agriculture; and
5. soil improvement opportunities.

Even where increasing amounts of hydrometeorological data have become available, their interpretation and evaluation has not been adequate, usually because the records cover too short a time span. Problems deserving close attention include methods of correlating rainfall and streamflow data; techniques for the estimation of floods; the analysis of low flow frequency; estimation of evaporation; and the analysis of climatic trends and the extent to which they may be relevant to overall planning of river development.

Information on the condition of catchment areas is not always available, although East Africa (Kenya, Tanzania, and Uganda) is rather lucky in this respect. The following information for catchment areas should be made available to planners:

1. water balance analysis;
2. effects of changes in land use on total water yield;
3. effects of changes in land use on sediment yield;
4. effects of changes in land use on seasonal distribution of streamflow;
5. development of rainfall-runoff models;
6. application of models to ungauged catchments;
7. estimation of actual evaporation; and
8. sedimentation rates in existing reservoirs.

Such data are at present badly lacking in most developing countries. In the final analysis though, the availability of data, no matter how adequate or inadequate, does not assume their correct use. As a result of the inadequacy of data, many plans for river basin development in developing countries are inflexible and rarely provide alternative strategies.

Land Use Planning and Its Application to River Basin Development

In order to achieve meaningful development in a river basin, the land use in the whole basin should be examined as a basis for putting forward alternative development proposals. The planned use of water, whether or not a scarce resource, inevitably involves the use of land. Proper land use planning, except for local projects, is yet to be fully employed in most developing countries. Land use studies that have been carried out often have been disconnected due to the fact that they were done by unrelated teams of foreign experts. Such studies may cover important portions of a river basin without any serious efforts to note that proposals for the development of one section would ruin the chances for development in another section. The Tana River basin in Kenya faced such problems in the past—now there are efforts to rationalize all future developments under one administrative structure.

Many studies for irrigation development within basins like the Tana have been carried out by firms of consulting engineers. These firms tend to concentrate on specific aspects designated by the individual governments. Only rarely do they adopt the multipurpose approach. The aim should be to produce complementary rather than divergent or conflicting development proposals. Ideal development planning involving a whole river basin should start by dividing the basin into units based on existing or potential land use characteristics. This can be followed by assessment of land and water use possibilities on the basis of detailed surveys of topography, soil, and vegetation.

The concept of land use planning brings with it the multipurpose approach to land and water use. For example, a dam constructed for power development can also be used for irrigation, fisheries, transportation, and recreation. What is wrong in most developing countries is that these multipurpose avenues have neither been properly explored nor fully utilized. The emphasis on land use studies within river basin systems could be the first move in the direction of more integrated planning.

Watershed Approaches to Multipurpose River Basin Development

The most logical units to use in such planning are the watersheds. The watershed is usually a clear-cut hydrological unit; it is usually an ecological unit; and it also may be important for a variety of human activities. In attempting to analyze the ecological and conservation implications of proposals to develop a river basin, the watershed approach can provide valuable clues of ecological and human impacts.

The main aim of most river basin development projects in Africa today is to increase food supply for the population within the basin or in the rest of the country. Thus, the main objective for development has been summarized as the need to increase the value of goods and services which the region is capable of producing for the benefit of the population. Usually a high value is placed on the short-term benefits to be obtained, but seeking this goal brings three types of problems.

First, one may encounter a situation of convergent factors within a given development area. In this kind of situation there are no major conflicts between optimizing output and income on the one hand and resource conservation and ecological balance on the other. Here, in order to achieve satisfactory and sustained levels of output, there must be the careful management of the existing resources. This makes conservation measures an essential and integral part of the development strategy.

A second situation is where the various factors, though divergent, are amenable to resolution. This is a common situation. The expectation that conflicts will occur in water usage, for example, is accompanied by confidence that they can be reconciled through a multipurpose approach. To minimize both foreseen and unforeseen conflicts, comprehensive planning prior to initiation of large scale investments is compulsory. The anticipated conflicts can be minimized in one of the following ways.

1. Modifying the original goals and targets. It is rare that the final version of a development project is ever the same as

when it was first envisaged. Flexibility is necessary in the interest of producing a realistic program.

2. Scaling down some activities and making others contingent upon prior action in other fields. For example, in Kenya there have been big proposals for the irrigation of the whole of the lower Tana basin. To avoid costly mistakes a decision has recently been made to start with a small project in the Bura area. The extension of the project to cover a much larger area will be contingent on success being achieved in the Bura project.

3. Development of a phased strategy of implementation in which each step is a logical precursor to the next. For example, in the Kano Plains area of western Kenya the expansion of irrigation to cover much of the region will depend on the success achieved in the Ahero project (called Phase I) and the west Kano project (called Kano Irrigation Phase II).

4. Allocating costs and benefits among the different social groups involved in a fair manner.

5. Arranging compensation for the losers.

Finally, there is the situation where the factors are seriously conflicting and where immediate solutions are not apparent. It is common to find proposed development measures in a river basin or in a watershed that so seriously endanger the whole ecosystem that the project should be regarded as being impractical. The dangers of irreversible ecological changes are only too apparent in many developing countries. The damage done may be much greater than the benefits accruing from the proposed development. Unfortunately, projects which are known to be ecologically unwise have been allowed to proceed for political reasons. (For example it appears that the proposed Jonglei Canal Project is likely to be executed by the Sudanese government despite many misgivings about its likely ecological impact in the southern part of that country.) The politicians ignore the inherent ecological risks such as the complete destruction of forest, irreversible soil damage, and rapid sedimentation in newly built dams. The best that the scientists or planners can do in such cases is to prepare a full inventory of alternative strategies which, in social accounting, may be less costly. Unhappily, the lending institutions and the bilateral aid agencies from the more developed world are often party to such environmental damage because they feel obligated to accept the political judgment of the national government as to the risks involved.

Watershed Management Problems

Careful study of watershed conditions may provide the planners in each of the situations noted above with alternative strategies for development which will be ecologically more acceptable, and which also may be technically sound and politically acceptable. These strategies are widely interpreted to cover natural as well as human settlement ecological conditions within the total area under consideration. Like the concept of a catchment area, they are not always well understood. In cases where there is lack of adequate data, and where there is a general ignorance about the detailed ecological conditions within a river basin proposed for development, the use of simulation models could provide some of the answers. Unfortunately, the conditions under which simulation models could be used in the absence of adequate and representative data are still either unknown or improperly understood.

It is well known that watershed conditions can exercise a decisive influence on the regime of a river. The nature and frequency of floods, low water discharges, and sediment transport all depend on the state of the watershed. The action of man in many watersheds exercises a decisive influence. Whereas erosion from a watershed is a normal, natural process, in most headwater catchments, erosion is multiplied many times by man and his animals. This is evident in East Africa where most rivers start from the high precipitation highland areas which become the focus of much human activity. Because these highlands are also the most attractive for human settlement and agricultural land use, increased and accelerated soil erosion is highly likely. Destructive cultivation methods, frequent forest fires, and overgrazing by livestock and wildlife combine to increase the rate of soil erosion from the watershed. The improvement of cultivation practices, maintenance of a good vegetation cover, and good forest management are the remedies which will help to maintain and restore a healthy watershed. But as experience from East Africa has shown, good watershed management cannot be considered only in terms of erosion control. It has side benefits such as the improvement of water regime by increasing low water discharges, and by reducing to some extent the flood peaks and the silt transport which normally choke the river beds in their downstream reaches. By reducing the silting of reservoirs, the useful lifetime of dams for power or for irrigation is extended, and the dams are made more economical.

Watershed management may lead, at the same time, to improved livelihood in the upper portions of basins. It is likely to increase agricultural production and allow more productive forest management. Where the headwaters of rivers discharge clean water into reservoirs, the development of fisheries is supported. But in order to achieve successful watershed management, cooperation between farmers and the goverment, and between farmers and other sectors of the community, is essential. In Kenya there has been much experimentation with soil conservation practices as part of the overall plan for watershed management. The lesson which is very apparent is that success can only be achieved where the individual farmers are prepared to cooperate with the government. This is true in densely settled high-potential land areas which are located in main catchment areas (Figure 8-1).

It should be emphasized that, since the effect of watershed management takes many years to realize, it is desirable that a start be made on proposed measures as early as possible. For example, the schemes which were established in Kenya immediately after the Second World War have only proved their true benefits in the last few years.[3] Thus, watershed management schemes should be started many years ahead of the construction of many engineering works on the main rivers and tributaries.

Some Lessons from Catchment Area Research in East Africa

East Africa has greatly benefited from catchment area research carried out by the East African Agricultural and Forestry Research Organization.[4] It was long realized that the

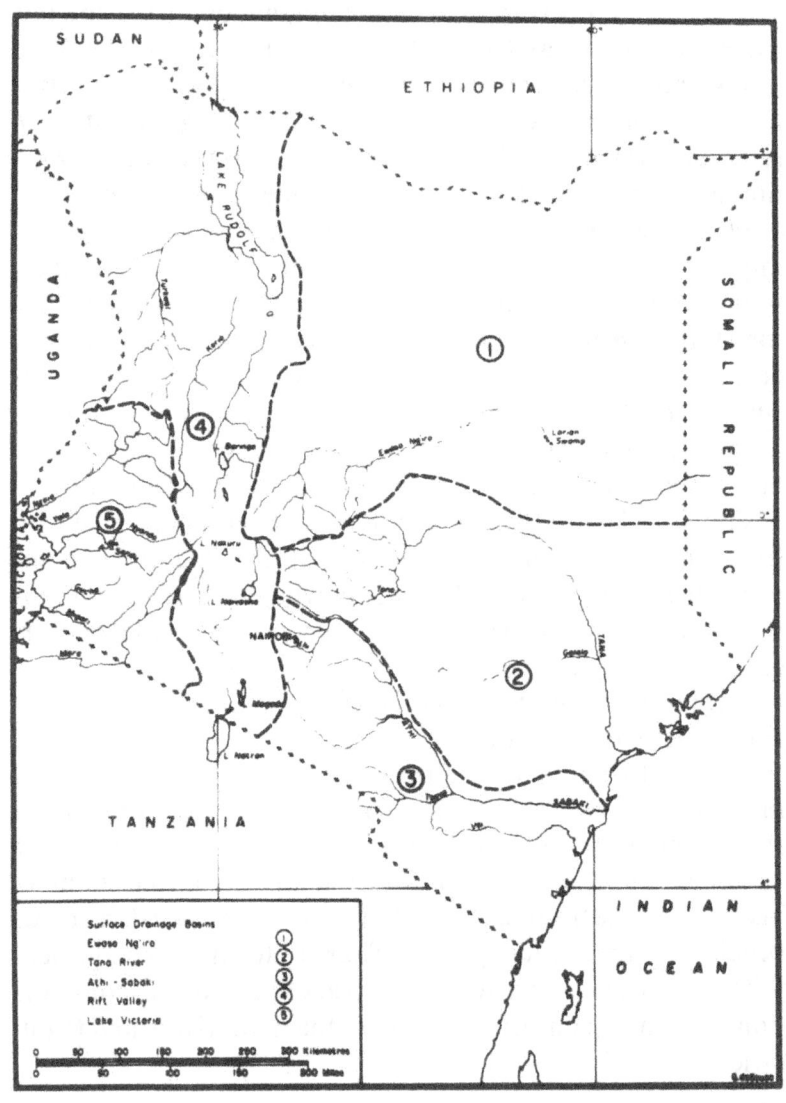

FIG. 8–1 THE DRAINAGE BASINS AND THE MAJOR RIVER BASINS
IN KENYA

high-potential land areas which were under gazetted forest reserves would one day be under pressure as population increased in East Africa, because these areas are also some of the most attractive from the agricultural point of view. To obtain scientific information on the response of catchment areas to human pressure, a few experiments were established. Two of these aimed at gauging the hydrological effects of possible land use changes in a major watershed.

Small sites were selected within certain major catchment areas. It was assumed that the results obtained from these sites would be applicable to the catchment as a whole. Although this assumption can be doubted on the grounds that each watershed is unique, the experimental approach has been justified on the grounds that the major effects of geology, topography, soil, climate, and vegetation characteristics are all reflected in streamflow, which can be accurately measured. Major changes in land use can be reflected in total water yield, as well as the pattern of river flow, but it may be difficult to separate out the influence of each factor contributing to the change. Wangati[5] has shown how these changes can be evaluated through comparison with streamflow of an adjacent, untreated watershed as follows:

The total water yield Q of a watershed is given by

$$Q = R - \Delta S - Et - \Delta G$$

where R = total rainfall received

ΔS = increase in water stored in the soil profile

Et = total water loss to the atmosphere through evaporation

ΔG = net ground water recharge

In this set of experiments, the groundwater recharge is the most difficult parameter to measure and every effort is made to select watersheds whose geological structure up to the point of measurement of streamflow does not permit significant water percolation to groundwater. The change in the soil profile moisture content S may be measured, and the time interval selected to minimize S. The experimental catchments should also be large enough to accommodate simulation of the real land use change with all resultant changes in human population and activities. Since evaporation loss is determined mainly by climatic factors, the factor Et is expected to stabilize to a constant ratio of potential evaporation E_0 once the new land use and ground cover is established.

One watershed was located in the high rainfall areas, which are the main forested areas in East Africa and which also contain the most agriculturally desirable land in the region. The danger in interfering with these watersheds lies in the possible removal of the regulating system, with resultant flooding of river channels during the rainy season, and reduced contribution to dry season flow.

The second watershed was located in the drier regions at medium to low altitude. Here the problem presented is that of shallow soils and high-intensity rain storms, while the total annual rainfall is less than potential evapotranspiration. Flash floods accompanied by high sediment transport and siltation rates are therefore fairly common.

In the case of the highland, forested catchment areas, it was observed that in their natural state the catchment areas were characterized by very little soil erosion, with surface runoff limited to about 3 percent of total rainfall. In changing such a watershed to pasture or tea plantation, it was found that evaporative loss was reduced from $0.8 E_0$ to $0.3 E_0$—this meant an increase in water yield approximately equivalent to some 900 m^3 per month for each hectare cleared.[6] Secondly, it was found that further land use changes involving human settlement, small-scale agriculture, road building, and grazing by livestock would rapidly accelerate soil and water loss from a catchment area to such an extent that it would be ruinous.

In the case of the lowland watersheds, it was found that even greater care is required. The low rainfall and the relatively poor and shallow soils make it difficult to establish rain-fed agriculture of even pasture grasses after the original vegetation is removed. In other words, land use changes in such catchments appear to be irreversible. The high rate of soil erosion from such a catchment not only reduces the recovery potential of the watershed, but it also reduces the life of dams or reservoirs because of the high and accelerated siltation rates.

The Ecological Impact of River Basin Development Projects

Where the demand for land leads to intensive human settlement in the catchment area, the resultant destruction to vegetation through clearing for cultivation, burning, cutting for charcoal making, and overgrazing has obvious consequences. However, it is possible to prevent such continued deterioration of the catchment area by a progressive agriculture and by application of strict conservation practices.

In ecological terms, the primary value of a river basin may be viewed as the total biomass it supports through its water volume. It has a system of food chains essential for the continued survival of its biomass (including man). Certain engineering measures executed by man in an effort to benefit a portion of the river basin, as in the case of an irrigation scheme, or hydroelectric power or urban water supply, may drastically change the total biomass or its composition. One of the common consequences of river basin development in developing countries is the total destruction of wildlife habitat, especially the riverine habitats, and its replacement with neat irrigation fields. Planners of river basin development are now becoming wary of projects which call for eventual destruction of wildlife habitats and wildlife altogether, because it has been shown in Kenya, as in other East African countries, that wildlife can be of great touristic value and thus serves one of multiple aims.

Although it is common to blame major engineering projects for ecological destruction in river basins, it has been shown in East Africa that certain simple technologies can also lead to destruction. For example, a sudden burst of charcoal burning activities in the upper and middle catchment areas of a river basin can strip the whole basin of its natural vegetation and its future biological carrying capacity.

To prevent such events it is necessary that prior to the

establishment of major development projects in a river basin, there should be surveys to provide planners with an overall inventory of primary value ecological constituents.

Following the establishment of projects there should be a constant ecological monitoring within the river basin in question to observe undesirable ecological effects which could be referred to as side effects of a major development program. In the same way as a balanced agricultural system in the catchment area of a river basin can help to curb rapid silting of reservoirs, so informed basin planning and monitoring can help to prevent other ecological disasters.

Human Ecological Considerations and River Basin Development

To underline the human impact of river development projects in developing countries, one can do no better than to quote the report of the Nobel Workshop on Schistosomiasis held in Stockholm in 1973:

> Water resources development schemes have been installed in large numbers especially in developing countries because of their obvious economic and social benefits and contributions to development. While many of the consequences, such as agricultural and economic, are positive and anticipated, many are not necessarily considered, and major consequences are negative. There is evidence that health is undermined by the projects due to the spread of waterborne diseases. Of these schistosomiasis is unquestionably a major factor.[7]

This workshop noted that schistosomiasis affects between 100 and 200 million people in seventy-one countries and that, unlike most other public health problems, this disease has been spreading and remains largely uncontrolled.[8]

Schistosomiasis represents one explosive type of disease situation created by river impoundments in developing countries—its importance is underlined by the fact that despite some control measures, it remains a disease much on the increase. This increase is partly due to manmade lakes being established for power, irrigation, municipal water schemes, and flood control, and partly due to vector snail infestation of new areas which have hitherto been free of schistosomiasis. Manmade lakes attract a lot of unsupervised settlements which are partly responsible for the spread of this disease.[9] Irrigation schemes in many developing countries like the Sudan (and Kenya to some extent) harbor these vector snails and also are responsible.

Apart from schistosomiasis, the construction of dams on rivers may lead to an explosive increase of malarial mosquitoes, and in some cases like the Volta Dam in Ghana, to the increase of the river-blindness fly *Simulium damnosum* and the consequent risk of Onchocerciasis.[10] Another common disease associated with water development projects in developing countries is sleeping sickness transmitted by tsetse flies, especially the species *Glosina palpalis*. In order to deal effectively with these and other waterborne diseases, it would help to employ transdisciplinary study teams to plan projects where the disease risk is appreciably minimized or altogether eliminated.

Agricultural and Livestock Rearing Implications

After waterpower, agricultural development is the most common justification given for interfering with river regimes in developing countries. A dam may be constructed across a major river to impound the waters and make it available for irrigation—as in the Tana basin. In another instance, swamp drainage can be followed by irrigation—as has been recently proposed for the Yala Swamp on the shores of Lake Victoria in western Kenya (Figure 8-2).

The Tana is the largest river in Kenya, with an irrigation potential of over 100,000 ha. Most of the irrigable land is found in the lower Tana basin, but in the middle basin, in the Mwea-Tebere Irrigation Scheme, some 6,000 ha of land are already under irrigation. This basin provides some interesting problems common to developing countries in the fields of planning, agriculture, and livestock management. Apart from the Mwea-Tebere Irrigation Scheme already mentioned, and some

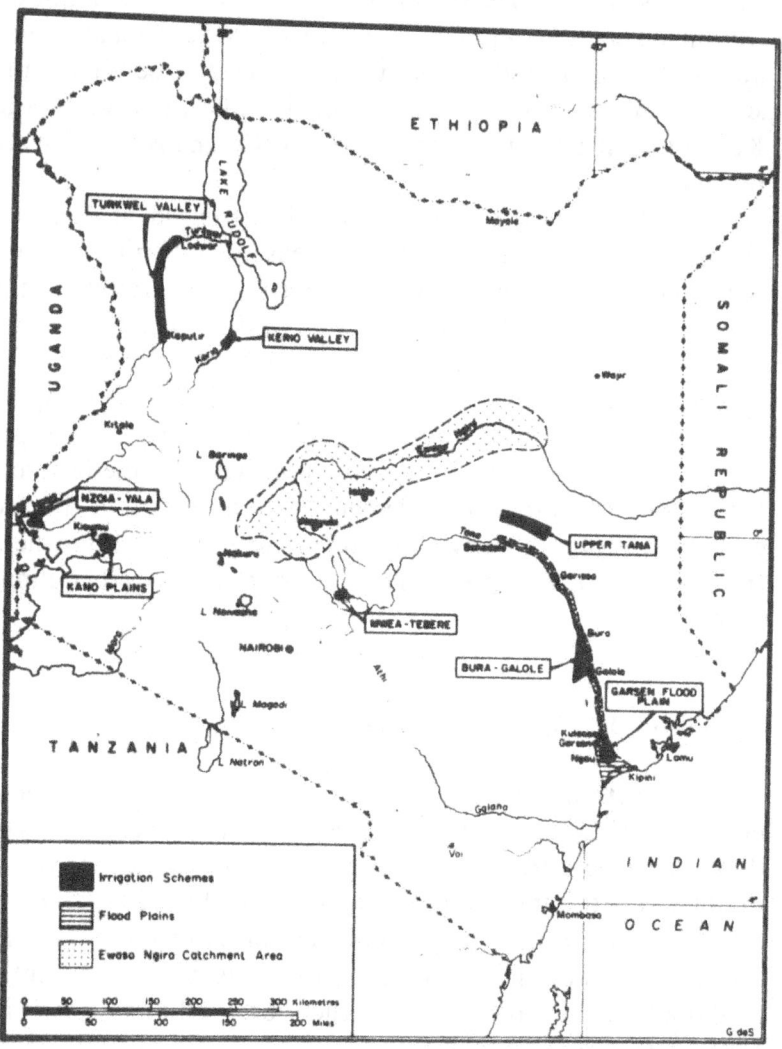

FIG. 8–2 KENYA: ACTUAL AND POTENTIAL IRRIGATION AREAS

smaller projects in the lower Tana, the agricultural systems found along this river basin are mainly subsistence-oriented. But in Kenya, as in many developing countries, agriculture is expected to grow sufficiently to feed the rapidly increasing urban and rural populations, and to provide raw material for agriculturally based industries. For this reason commercially oriented agricultural development is an attractive proposition, and an area with such large potential as the Tana River basin becomes a natural focus for future development proposals.[11] But such capital-intensive irrigation projects as have been under consideration for the Tana are fast becoming unrealistic. The costs are heavy: it would require over K£200 million to develop the existing irrigation potential in the lower Tana basin (average annual national income is K£674 million). The success of such irrigation schemes would largely depend on the availability of sufficient water for irrigation and the existence of a carefully integrated development plan covering the whole river basin. This, unfortunately, is just what is lacking in the case of the Tana basin.

The basin demonstrates the piecemeal approach. During the colonial period the Mwea irrigation scheme was planned and commenced.[12] Soon after independence, plans for a hydro-electric complex involving the construction of three power dams (Seven Forks Complex) at Kindaruma, Kamburu, and Gtaru, were laid down.[13] More recently, a detailed study of the irrigation potential of the lower Tana Valley was carried out by a team of irrigation planning experts.[14] No adequate attempts were made to integrate these studies even though it was known that some of the development proposals would obviously conflict. The earlier studies were financed by different bodies and were envisaged as single-purpose projects. Recently, however, the Kenyan government decided to create an overall administrative authority for the whole river basin—the Tana River Development Authority. Hopefully, it will foster an integrated approach and bring about a harmonious development in the whole river basin which will benefit agriculture, waterpower development, transportation, fisheries, and tourism.

The Mwea-Tebere Irrigation Scheme in the middle Tana is considered to be one of the most successful irrigation schemes in Kenya.[15] Much of the land preparation and leveling was done during the colonial period using convict labor. Water for irrigation is derived from two upper tributaries of the Tana— the Thiba and Nyamindi—with limited catchment areas of 224 and 176 km^2 respectively. The total area under irrigation is approximately 5,000 ha, out of a possible maximum of some 6,000 ha (Figure 8-2). Rice is the only crop grown under irrigation, and it is grown on both black and red soils. The black soils give little difficulty in irrigation while the red soils, because of high infiltration rates, present a major problem. The whole scheme is divided into numerous tenant holdings among some 3,000 tenants, giving an average holding per tenant family of 1.6 ha. The whole scheme, as managed by the National Irrigation Board, provides a livelihood for some 30,000 people annually. The average yield of rice, which amounts to 75 bags (75 kg each) per hectare, is regarded as being extremely high and the average income of each tenant family averages some 3,497 shillings (K£175). This small irrigation scheme accounts for up to 70 percent of Kenya's rice demand. Although the scheme is relatively free of problems, the annual crop damage by Quelea birds *(Quelea quelea)* is fairly serious. Schistosomiasis is present, and all the main canal intakes are treated with molluscicides to keep down the vector snail population.

The Mwea-Tebere Irrigation Scheme demonstrates the advantages of stepwise development and relatively small project size. Management is fairly effective. It was constructed using simple and inexpensive structures, which have proved easy to operate. One major advantage is that almost all irrigation is by gravity, thereby drastically cutting the cost of moving water to the fields. All the supporting services— including the supply of seeds, fertilizers, service roads, credit, and marketing—are built into the scheme. Its success and profitability are not often repeated in many other schemes in developing countries. The experiences gained in this pioneer scheme can now be used as the basis for planning in the other, much larger, schemes projected for the lower Tana.

Livestock Raising Problems in the Lower Tana

The environmental problems likely to be encountered in the lower Tana may be suggested by reviewing the implications of such development for livestock raising in the area. The lower Tana is in an arid and semiarid region of Kenya (Figure 8-3). Before reaching the humid coastlands the river flows through an arid belt mainly used by wildlife and nomadic pastoralists as rangelands. Much of this area is tsetse infested. The population moves back and forth from the river valley according to the seasons. The dry-period grazing areas are largely concentrated in the alluvial plains and bottomlands close to the river. During the wet season, these areas are flooded and the cattlemen move to higher grazing lands. It is in this area that proposals are made for irrigation schemes which could ultimately cover an area of some 100,000 ha.

The area under consideration varies in elevation from 180 to 480 m above sea level. Rainfall totals range from 600 mm per year on the wetter fringes to as low as 300 mm on the drier fringes. Such rainfall as falls in the whole area is variable in occurrence from one year to the next, and from season to season. Livestock production in the area depends on the little rainfall that is received and on the presence of a permanent watering place, namely the River Tana. Proposals for irrigation could lead to the complete displacement of the pastoral population. Most planners in similar situations regard it as axiomatic that when irrigation comes, the nomadic populations must be settled, and must be prepared to give up their lifestyles and become sedentary cultivators. This may, however, be contrary to their wishes and lead to the failure of the irrigation scheme as a whole.[16] Quite apart from the problem of trying to sedentarize the nomadic pastoralists, the introduction of irrigation would lead to the decline of livestock production.

In a dry area like the lower Tana, the basic problem facing the livestock producer is the carrying capacity of the land, in terms of livestock as well as the wildlife, which depends on the natural herbage. The carrying capacity may be expressed as the livestock and wildlife biomass which can be supported during the wet and dry seasons. It is common to find such areas

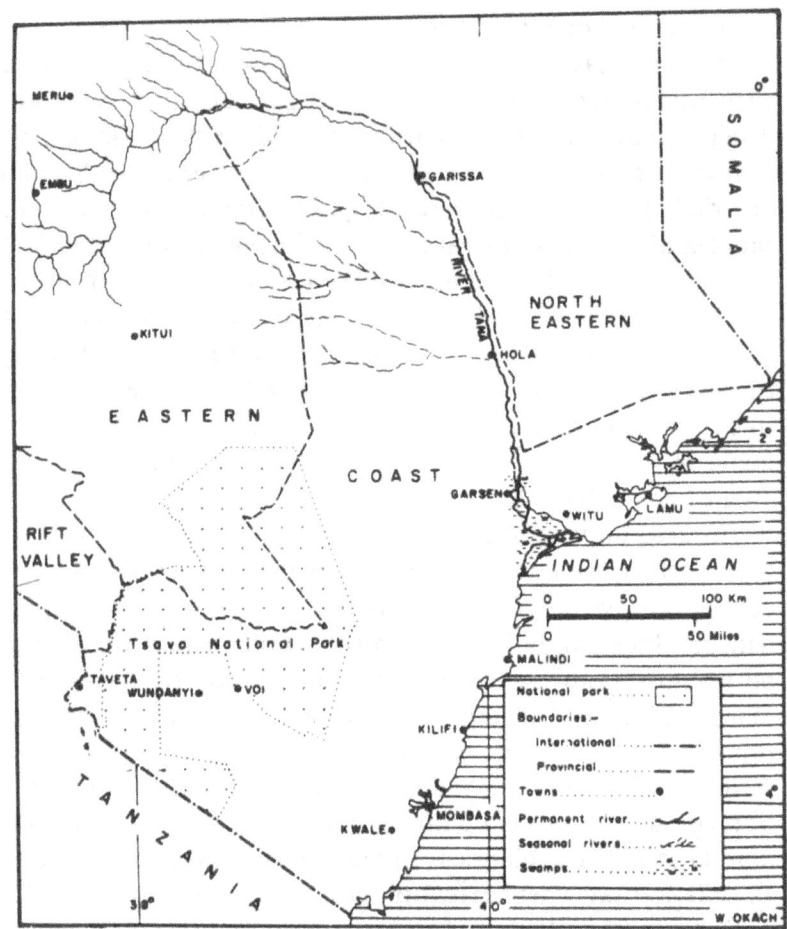

FIG. 8–3 THE TANA RIVER BASIN IN KENYA

overgrazed and the natural vegetation much destroyed by overstocking and consequent overuse. Vegetation which is subjected to excessive grazing by game and livestock is bound to deteriorate. This is particularly true of the area within 10 km of the Tana River today. Quite often the flood-induced grass-lands in deltaic areas or in inland swamp areas are of cardinal importance to the survival of the game livestock complex. A common range management practice recommends the use of flood-induced grazing areas during the dry season and other grazing areas during the wet season. The nomadic herdsmen who live in this place are the Orma, Boran, and Somali, who keep cattle, camels, goats, and sheep. They follow the rains with their livestock during the rainy season and gravitate back to the river during the dry season. The taking over of the whole area for irrigation would seriously disrupt this pattern of life in many ways. The traditional dry season grazing areas would be replaced with irrigated fields. The pastoralists would lose access to river water during the dry season, and this would become a source of conflict. Cost-benefit studies, carried out by range management specialists, have shown that it would be uneconomic to use the Tana River as a potential source for piped water to supply the rangelands in the interior which would be cut off by the arrival of the irrigation project. Planners will need to find alternative solutions to the problems posed. For example, plans for the irrigation scheme should incorporate proposals for irrigated fodder as a substitute for the loss of the dry-season grazing areas.

One of the major characteristics of the lower Tana basin is that in addition to supporting a large group of nomadic pastoralists and some cultivators (the Pokomos, who occupy the riverine flood plains), it also happens to support a rich wildlife biomass which ranges from small mammals to ele-phants, rhino, topi, oryx, etc. Most of these plains game depend on the river for grazing during the dry season, and on the river for watering most of the time. Any river development which destroys this wildlife heritage would be unfortunate. The planners should therefore make provisions in their multipur-pose approach for the preservation of this wildlife.

The Tana River clearly demonstrates the problems which many developing countries face in their efforts to bring about developments in a river basin. Integrated planning can anticipate many of these problems and provide solutions through alternative development strategies. This would involve measures such as provision of acceptable alternative dry-season grazing areas for the nomads, the incorporation of fodder growing in the newly irrigated lands, the provision of alternative sources of watering for cattle during the dry season, and carefully planned resettlement schemes which would minimize the ecological disturbances of the river basin. Many planners charged with development plans in developing countries have not always been sufficiently flexible and comprehensive in their approach. They have therefore failed to produce harmonious development in spite of the large sums of money spent.

Demographic and Human Settlement Considerations

In writing about the state of dam building as witnessed in many African countries, Pereira has the following to say: "Population growth, compounded by the great increase in the consumption of water per head by a modern technological society, has caused a vast increase in the rate of man's dam building efforts in the past fifty years."[17] In the context of multipurpose river basin development in developing countries, man is important both as a user and a resource. The problems faced by African countries in the course of river basin development will depend on the demographic characteristics of the population, including size, structure, growth and growth components, distribution, movement, and settlement pattern, within and without the development area under consideration.[18] The decision to develop a river basin for irrigation purposes is often seen in the context of the growth of population vis-à-vis the growth of agricultural production—food production as well as the production of industrial crops. In Kenya this problem is very apparent. The high-potential land areas located in the highlands are virtually saturated today by population. High rural densities are apparent everywhere. Planners are bound to look for alternative areas, which can be developed agriculturally in order to absorb some of the surplus population, provide new employment possibilities, and produce more food. The sparse population in the lower Tana Valley is seen as a hindrance to development. The growth of population leads to mobility in search for new settlement opportunities as well as employment possibilities. This explains why a newly built dam like the Kamburu Dam in eastern Kenya is suddenly undertaken, and by and large unsupervised and unplanned for settlement.[19] Thus, population mobility becomes an important aspect of river basin development. It is common to find population attracted to the vicinity of new development. In other cases, a manmade lake results in the displacement of population. The best example of this is found in the Volta Dam area in Ghana where a population of more than 80,000 was resettled. Settlement of displaced population constitutes one of the more delicate aspects of modernization. Planning for resettlement populations usually suffers from the total lack of information as to the future size of the population affected and changes in the various functional groups within the population.

The population problem is best seen in the catchment areas where pressure on land quickly leads to the upsetting of the delicate ecological balance associated with these areas. The consequences of misuse of land due to population pressure in these areas affects all the downstream users of the river. A rapid increase in the rural population makes it extremely difficult to apply soil conservation practices suggested for users of land. As land becomes more scarce, new resource management problems arise. The increasing pressure of population in the catchment areas must be regarded as one of the crucial development problems. Erosion of the headwaters will continue so long as populations of subsistence agriculturalists and grazers increase.[20]

There is already evidence of competition for land and water between urban and rural users. This is particularly true since the decision to establish numerous rural water reticulation programs. The likely effect is that there will not be enough water for irrigation and power requirements as well as for urban and rural requirements. Integrated development of a river basin therefore calls for a rational apportionment of the scarce water resource and, even more important, an accurate assessment of the competing needs. Integrated development should anticipate the requirements of the new settlements, their share of the scarce water resources, and of land and energy.

The most dramatic symptoms of land misuse are floods, although not all floods are due to man's mismanagement. These problems are compounded by the lack of effective legislation to stop cutting down trees for charcoal burning. The effects of such practices on river regimes are disastrous.

Institutional Considerations

In a developing country like Kenya, the management problems are usually more critical than the unavailability of resources. River basin development has in the past been loosely left in the hands of civil engineering firms and aid experts. Specific activities requiring more specialized management come under statutory boards like the National Irrigation Board. It is only recently that many developing countries have begun to realize that effective management of integrated river basin development requires more powerful institutions. The details of the kind of structure which should be used will depend upon the special circumstances of each country. In any case, to iron out problems like: lack of coordination at all levels; imprecise development goals; inadequate laws and law enforcement; lack of skills to formulate, execute, and manage sound projects; and lack of public support for multidisciplinary approach to conservation of the environment, a strong central authority is necessary. Some observers argue that a river basin authority should be used.

In Kenya, the realization of these problems led to the creation of the Tana River Development Authority. A river basin authority offers a promising possibility for proper administration provided it is vested with the proper powers. The authority should regulate development according to a master plan which aims at integrated development and which includes all aspects of human welfare. Since a few river basin authorities have proved successful in the developed world,

African countries can borrow from their experience and tailor to their particular needs. Unfortunately, there is still a lot of reluctance to create powerful institutions for development in most African countries, and indeed in many developing countries.

Summary and Conclusions

In the foregoing paragraphs, different aspects of river basin development have been examined with a special focus on the type of problems which are being encountered or are likely to be encountered in developing countries. The lack of sufficient and reliable data has hindered proper integrated river basin planning. However, problems often arise because of lack of sufficient interest or ignorance of the best way to develop a river basin. Information systems about catchment areas can be greatly improved by catchment area research intended to provide specific answers to specific questions. The planning of dams for irrigation and for power production should, as much as possible, be multipurpose, as this will enable the people served to reap the maximum benefits of the development. Developments which bring about ecological damage or total ecological disruption should be avoided. Finally, every effort should be made to develop institutions which are strong enough to bring about unified development in whole river basins.

Notes

1. Kenya, *Development Plan 1974-78* (Nairobi: Government Printer, 1974).
2. L. E. Obeng, "Integrated River Basin Development," in *Environment and Development,* edited by D. M. Dworkin (Indianapolis, Ind.: SCOPE Miscellaneous Publications, 1974).
3. Kenya, *African Land Development in Kenya* (Nairobi: Government Printer, 1962).

4. *East African Agriculture and Forestry Journal,* Special Issue on Catchment Area Research in East Africa (1962).

5. F. Wangati, "Land Use Changes in East Africa," paper presented to the Workshop on Problems of Multi-purpose River Basin Development in Developing Countries (Nairobi: September-October, 1975).

6. F. Wangati, "Land Use Changes in East Africa."

7. IFIAS, *A Plan for a Transdisciplinary Study of Events Occurring when a Tropical Water Development Scheme is Implemented, with Special Regard to its Bearing on Shistosomiasis,* Report from the Nobel Workshop (Stockholm: Nobel Foundation, 1973).

8. IFIAS, *Plan for a Transdisciplinary Study.*

9. SCOPE, *Man-Made Lakes as Modified Ecosystems,* SCOPE Report 2, International Council of Scientific Unions (1972).

10. Food and Agriculture Organization, *Health Implications of Water Related Parasitic Diseases in Water Development Schemes* (Rome: FAO, 1967).

11. Kenya, *Development Plan 1974-78.*

12. Kenya, *African Land Development in Kenya.*

13. R. S. Odingo (ed.), *Kamburu-Gtarv Ecological Survey: A Preliminary Report* (Nairobi: University of Nairobi, 1975).

14. Kenya, *Survey of Lower Tana Basin* (Nairobi: Government Printer, 1974).

15. R. Chambers and J. Moris, *Mwea-Tebere Irrigation Scheme* (Munchen: Weltform Verlag, 1973).

16. W. Allan, *The African Husbandman* (London: Oliver and Boyd, 1965).

17. H. C. Pereira, *Land Use and Water Resources* (New York: Cambridge University Press, 1973).

18. S. H. Ominde, "Demographic Aspects of Multi-purpose River Basin Development," Paper presented to the Workshop on Problems of Multipurpose River Basin Development in Developing Countries (Nairobi: September-October 1975).

19. Odingo, *Kamburu-Gtarv Ecological Study.*

20. Pereira, *Land Use and Water Resources.*